AF811550

Ham Radio General Class Study Guide 2025-2026

Complete Review + 210 Questions and Detailed Answer Explanations (6 Full-Length Exams)

Printed in the United States of America

Table of Contents

****There is a QR code on the last page of this book that can be used to download the audiobook. Please contact info@newstonetestprep.com if you are unable to download the file and mention the title of this book.**

Introduction

About the Ham Radio General Class License Exam

The General Class license is the second tier in the United States amateur radio licensing structure. It gives you access to a more comprehensive range of HF (shortwave) bands and enables communication across the country and around the world.

Structure of the Exam

The Ham Radio General Class exam comprises 35 multiple-choice questions, each with four possible answers. To pass the exam, candidates must correctly answer at least 26 questions, which is 74% of the total. The 35 questions are drawn from the following amateur radio concepts:

- FCC Rules (5 questions).
- Operating Procedures (5 questions).
- Radio Wave Propagation (3 questions).
- Amateur Radio Practices (5 questions).
- Electrical Principles (3 questions).
- Circuit Components (2 questions).
- Practical Circuits (3 questions).
- Signals and Emissions (3 questions).
- Antennas and Feed Lines (4 questions).
- Electrical and RF Safety (2 questions).

Why General Class is Worth the Effort

Expanded Privileges

The General Class license expands the frequencies and modes you can transmit and receive. The table below illustrates the benefits of a General Class license compared to a Technician Class.

Feature	Technician Class	General Class	Notes
HF Bands (voice & data)	Limited access (10-meter band).	Full access (80-, 40-, 20-, 15-, and 10-meter bands).	Opens a vast range of global contacts.
Digital Modes (FT8, PSK31)	None.	Full access.	Allows experiments with data transmission and weak-signal communication.
Satellites	Uplink on select satellites.	Expanded uplink/downlink privileges.	Facilitates communication through space-based repeaters.
Contesting	Limited portions of HF bands.	Access to the full range of HF contest segments.	Allows a person to test their skills and make more contacts.

Community and Growth

The General Class license unlocks a world of connection and growth within the ham radio community. This vast network of fellow enthusiasts broadens your horizons.

Special interest groups (SIGs) help you to dive deeper and explore niches aligned with your interests. The American Radio Relay League (ARRL), for example, sponsors a wide range of affiliated SIGs that cover topics from contests and digital modes to QRP (low-power) operations. A quick web search will reveal numerous niche groups dedicated to vintage equipment, satellite communication, and emergency preparedness.

DX-peditions provide the thrill of the chase. You can track major DX-peditions online, where many provide live updates, logs, and even audio streams.

Hams can participate in formal or informal nets, from daily check-in nets to specialized traffic nets vital for emergency response. You can monitor local repeaters and HF bands or use online sources like the ARRL Net Directory.

Contests are open to all experience levels, from major international events to smaller regional ones. The ARRL Contest Calendar and dedicated contest websites provide valuable resources, rules, and event information.

The ham radio community extends far beyond the airwaves. Hams can connect with like-minded individuals at local clubs. You can use the ARRL Club Locator to find clubs in your area. Also, mark your calendar for regional conventions and ham fests. They are excellent venues to browse gear, listen to informative talks, and socialize.

Discovery

Imagine a world beyond line-of-sight communication, where radio waves dance with the ionosphere, and the very atmosphere shapes your signal's journey.

- **Global Connections**: There is an undeniable thrill when a faint signal resolves into a conversation with someone thousands of miles away. One day, you might chat with a fellow ham in South Africa. The next, you could talk with a station in Japan. With each contact, you forge a connection that transcends borders and reminds you of the global community you now belong to.

- **Next-Level Tech**: The HF bands are a hotbed of experimentation. Discover the world of digital modes, where your computer and radio collaborate to send signals that defy noisy conditions. You can explore software-defined radio (SDR), where limitless possibilities replace the traditional "black box" of radio.

- **The DIY Spirit**: Amateur radio allows you to design and construct antennas to maximize your signal. Ham radio also introduces you to propagation science and helps you understand how solar flares and atmospheric patterns affect your reach. This blend of hands-on experimentation and science is at the heart of ham radio.

What to Expect on Exam Day

Your first step is to find an exam session. The ARRL Exam Session Finder is the best place to start your search. You can also contact local ham radio clubs in your area, as they often host exam sessions.

Once you have found an exam, ensure you understand the registration process. Some VECs (volunteer examiner coordinators) require you to preregister. Confirm the exam fee, the exact test location, and any additional instructions the VEC may have.

On exam day, bring these essential items: a government-issued photo ID, your FCC Registration Number (FRN) if you have one, FCC Form 605 (only if you do not have an FRN), pencils, and a basic memory-cleared calculator with no stored formulas. Optional items like scratch paper, comfortable clothes, and a water bottle can also be helpful. Listen carefully to the examiners' instructions and manage your time wisely during the exam. If you get stuck on a difficult question, do not panic. Move on and return to it later.

After the exam, you may receive your results immediately. If you pass, the examiners will help you complete the necessary paperwork for the FCC. The FCC will then take some time to process your license application and assign your call sign, which will usually take a few weeks.

Overcome Common Obstacles

The journey toward your General Class license might present a few challenges, which is normal. The best approach is to acknowledge these hurdles head-on and approach them methodically.

Fear of Math

If the thought of equations induces anxiety in you, this is not unusual. However, the General exam, although it includes calculations, does not necessitate advanced mathematical expertise. Consider the calculator a reliable ally—its use is permitted during the exam. Understand the rationale behind formulas and recognize when to implement them. It is essential to prioritize the application of knowledge over mere memorization.

Technical Jargon

The FCC question pool can occasionally resemble a cryptic foreign language. However, there is no need for concern. A comprehensive glossary can help you decipher technical terminology.

Master the Regulations

A successful General Class candidate requires skills beyond mathematical proficiency. A responsible ham radio operator must be familiar with the regulations that maintain safety and order on the airwaves. Established procedures exist to facilitate seamless communication. Mastery of these guidelines will enhance both your confidence and effectiveness on the air.

Practice Spaced Repetition

Our brains naturally tend to forget information over time. Spaced repetition is a scientifically proven strategy to combat this. Don't cram right before the exam. Instead, revisit the material at intervals. You water a plant regularly, not all at once, or you might drown it. The same thing applies to how you learn—if you learn in increments, you're more likely to remember than if you flood your brain all at once with abstract concepts.

Resources: Your Study Toolkit

The right tools are essential for any exam candidate. The core resources for your General Class journey include this study guide, the FCC question pool, and any additional websites or texts you deem useful.

Let's Get Started!

The General Class license unlocks a vast world within ham radio. Although it requires dedication and effort, the rewards when you obtain a General Class License are immense. To set yourself up for success, try to understand the challenges, adopt effective study strategies, and embrace the journey with a growth mindset.

Chapter 1: Commission's Rules

Amateur radio operators must stay informed about the rules that oversee their broadcasts, especially on HF bands, where their signals have the potential to traverse continents. Organizations like the Federal Communications Commission (FCC) set guidelines to promote the secure and considerate operation of amateur radio wavelengths. Just like drivers need to follow the rules of the road, ham radio operators have guidelines. This section will discuss the agencies that make the rules and what is expected from ham operators to keep the hobby fun and legal.

International Regulatory Agencies

The global operation of amateur radio is regulated and facilitated by international organizations, mainly the International Telecommunication Union (ITU) and the International Amateur Radio Union (IARU). The ITU, a UN agency, governs spectrum allocation and sets technical standards, while the IARU represents the interests of amateur radio operators at the ITU and other international bodies. Together, they ensure a coordinated and standardized approach to amateur radio regulations worldwide.

International Telecommunications Union (ITU)

The ITU is a United Nations agency that governs global telecommunications, which includes the amateur radio service. It allocates radio spectrum, establishes international technical standards, and convenes the World Radiocommunication Conference (WRC). The ITU designates specific frequency bands for amateur use and ensures operators have dedicated space for communication, experimentation, and disaster preparedness. The ITU's radio regulations provide standardized guidelines for modes of transmission, power limits, and identification protocols. They also promote compatibility and seamless communication between amateur stations across the globe.

Notably, the ITU establishes the framework for international radio regulations, which individual nations implement through national laws and regulations, often formalized by treaties. Each country retains the right to administer and supplement those regulations within its jurisdiction, provided there are no conflicts with ITU standards.

To manage regional variations, the ITU divides the world into three administrative regions, each with its own set of frequency allocations. North and South America, Alaska, Hawaii, and most US territories fall within Region 2. These regional divisions impact frequency allocations for amateur operators. For instance, the 75-meter allocation varies across ITU regions, and further variations exist between countries.

International Amateur Radio Union (IARU)

IARU is a global coalition of various national amateur radio societies. Since its inception in 1925, it has been the collective spokesperson for ham radio enthusiasts worldwide.

Mission

The IARU's main job is to be the voice of ham radio enthusiasts at the ITU and other regional organizations. Its mission is to promote and grow amateur radio, protect amateur frequency bands from encroachment and align regulations across borders.

Membership and Structure

The IARU welcomes national amateur radio societies as members and currently numbers over 160 member societies in many countries and territories. The organizational structure of the IARU is divided into three distinct regional bodies:

- **IARU Region 1** encompasses the geographical areas of Europe, Africa, the Middle East, and Northern Asia.
- **IARU Region 2** represents the Americas.
- **IARU Region 3** covers the Asia-Pacific region.

The management of the IARU is governed by an administrative council, which consists of representatives from each of the aforementioned regional bodies. At the same time, an international secretariat carries out the day-to-day operational oversight and coordination.

The IARU holds the official designation of "Sector Member" within the ITU. This title grants it the authority to help formulate and regulate radio-related policies and guidelines.

In addition to its regulatory responsibilities, the IARU actively advocates for and promotes the use of amateur radio as a communication resource during emergencies and natural disasters. To this end, the organization provides comprehensive training and preparedness programs for amateur radio operators. It also equips operators with the necessary skills and knowledge to serve as effective communication support personnel in the event of infrastructure failure or disruption of traditional communication systems.

National Regulatory Agencies in the United States

In the United States, amateur radio operations are regulated and overseen by multiple national agencies, each with a distinct role. The FCC sets the legal framework, the ARRL advocates for amateur radio interests, and the FCC ensures regulatory oversight.

Federal Communications Commission

The FCC regulates amateur radio operations in the United States. The rules and guidelines for amateur radio practice are established in Title 47 of the Code of Federal Regulations (CFR), Part 97. This section of the CFR contains a comprehensive set of regulations that cover license classifications and privileges, examination criteria, technical specifications, and operational standards. It also caters to emergency communications protocols and prohibited practices within the amateur radio service.

Part 97 is the law for any ham in the United States. You can find all you need to know about Part 97 online at the e-CFR website, courtesy of the US Government Publishing Office.

American Radio Relay League

The ARRL is a nonprofit, membership-based organization established in 1914. It strongly advocates for amateur radio operators and represents their interests before the FCC and other regulatory bodies. It also lobbies to protect and expand amateur radio frequency allocations. The ARRL demonstrates the service's value to the government and the public.

In addition to advocacy, the ARRL also provides educational resources. It publishes the *QST* magazine, a monthly periodical that provides members with technical articles, operational guidance, and relevant amateur radio news. The organization has a library of books and study guides about various amateur radio topics to support individuals who want to obtain a license or enhance their knowledge base.

Federal Aviation Administration

The FCC is the regulatory authority that oversees the licensing and operation of amateur radio towers or antennas, but the FAA has specific regulations regarding towers or antennas that exceed certain heights or are located near airports.

The FAA conducts comprehensive aeronautical studies to determine whether a proposed structure could obstruct air traffic operations. These studies are a mandatory requirement for amateur radio towers and antennas to ensure a thorough assessment of their impact on the airspace environment.

Who Enforces the Regulations?

The regulatory oversight of ham radio operations has undergone a transition over time. Previously, this responsibility was undertaken by designated field officers appointed by the FCC. However, due to budgetary constraints, the amateur radio community has now adopted a self-policed approach to ensure compliance with established regulations. This paradigm shift toward self-regulation is facilitated by a specialized group of amateur radio operators known as volunteer monitors.

Volunteer Monitors (VMs)

The amateur service is a self-regulated community that complies with FCC regulations with minimal government intervention. The ARRL established the Official Observer program to assist the FCC with enforcement and interference issues. This program has now evolved into the Volunteer Monitoring Program (VMP).

The VMP promotes better compliance with FCC rules and broader knowledge of the regulations through a structured framework for volunteer engagement. It supports the concept of self-regulation and compliance among amateur radio operators.

Individuals are trained and screened by the ARRL to monitor radio frequencies to gather evidence in cases of infractions. This evidence is then reported to the FCC by the ARRL for appropriate action based on FCC protocols.

Amateur Radio Licensing Regulations

To obtain an amateur radio license in the United States, there is a structured process overseen by the FCC and facilitated by volunteer examiner coordinators (VECs) and volunteer examiners (VEs). This process includes specific examination, license generation, and renewal guidelines that ensure that all operators meet the necessary qualifications and adhere to established regulations. Beyond licensing, additional regulations govern amateur radio operation, from tower installation to remote control usage.

Volunteer Examiner Coordinators (VECs)

The FCC was in charge of all aspects of licensing until 1984. In 1984, VECs took over the licensing program. VECs are accredited organizations authorized by the FCC to oversee the administration of amateur radio license examinations within the United States. These entities develop and maintain exam question pools. They also recruit and manage VEs, facilitate exam sessions, and handle applications and results. The VEC system maintains a decentralized and efficient process to obtain and upgrade amateur radio licenses.

Volunteer Examiners (VE)

VEs are licensed amateur radio operators that administer amateur radio license examinations. Interested operators must possess a General Class or higher FCC-issued license to serve in this role. They must also obtain accreditation through a VEC. A VE must hold a license class above the exam to be administered and be at least 18 years of age.

Non-US citizens can become a VE, provided they meet the following requirements:

- Hold a valid FCC-issued amateur radio license (General Class or higher).
- Are at least 18 years old.
- Obtain accreditation through a VEC.

Examination Guidelines

Amateur radio exam sessions adhere to standardized rules outlined in Section 97.509 of the FCC regulations, regardless of the specific license elements that are tested. Each session must be coordinated by a VEC and overseen by at least three VEs accredited by the VEC.

Successful candidates receive a Certificate of Successful Completion of Examination (CSCE). This document indicates the elements passed and remains valid for a year. The CSCE is not mandatory documentation when one applies for a new amateur radio license but is required for upgrades to a higher class. Until the FCC issues a new license, amateur radio operators must continue to use existing call signs in conjunction with the CSCE.

Expired Licenses

Amateur radio licenses have a 10-year term. When a license expires, the FCC grants a two-year grace period. The license may be renewed without an examination during this grace period. However, you are not permitted to transmit on any amateur radio frequencies with an expired license, even within the grace period.

To renew an expired license within the grace period, you must complete FCC Form 605 and submit it with any applicable fees. If your license remains expired after the two-year grace period, your call sign will be forfeited and could be reassigned. In this case, you must pass the required exams for your desired license class to regain your amateur radio privileges.

The FCC has a streamlined reinstatement process for individuals whose Advanced or Extra Class license has expired. If you previously held one of these higher license classes and your license was not revoked, you may be eligible to regain your operating privileges without the need to retake the complete set of exams.

To qualify, you must provide evidence of your previous license and confirm that the FCC did not revoke it. Once you meet these criteria, you must pass the current Technician Class (Element 2) written exam and submit the appropriate documentation to the FCC or your VEC.

This process recognizes your prior technical knowledge but still ensures proficiency in core radio regulations, radio operating procedures, and basic electronics.

The FCC's Universal Licensing System (ULS) allows you to check your license status and manage your renewal process.

What is /AG?

The "/AG" suffix is a temporary designation for amateur radio operators who have successfully passed the General Class license exam but haven't had their license issued by the FCC yet. Once they pass the exam, ham operators gain immediate access to the expanded frequencies allocated for General Class license holders (including portions of the HF bands).

During the period that hams hold the /AG designation, they can still operate within the Technician Class portions of the bands, but they are not required to use the /AG suffix in those segments. When you transmit on General Class frequencies, you must identify yourself with your standard callsign followed by a slash and the letters "AG." For example, if your call sign is D1ABC, you would identify yourself as "D1ABC/AG."

Additional Regulations

Amateur radio operations are subject to additional regulations for antenna installation and remote control usage. Antenna installations may face local zoning ordinances or HOA restrictions, but the FCC's PRB-1 ruling mandates reasonable accommodation for amateur radio setups. Remote control operation is permitted, but regulations vary based on the transmitter's location, and restrictions on third-party traffic always apply.

Tower Installations

Antenna installations often face restrictions beyond those imposed by the FAA. These restrictions can be divided into two categories:

- Local government zoning ordinances.
- Covenants, Conditions, and Restrictions (CC&Rs).

Municipalities and local governments have zoning ordinances that regulate antenna structures' height, placement, and aesthetics. Homeowners in communities with a homeowners association (HOA) may be subject to CC&Rs. These legal documents often contain provisions related to antennas and towers, which may include outright prohibitions or strict limitations.

Amateur radio operators who face restrictive local ordinances or CC&Rs must know about the FCC Preemptive Ruling (PRB-1). This requires local governments and HOAs to "reasonably accommodate" amateur radio installations. The emphasis is on the technical requirements needed for effective amateur radio communication. If a proposed antenna installation conflicts with local zoning or CC&Rs, you must present technical justification for the structure's height or placement. Amateur radio operators are encouraged to engage in constructive dialogue with local authorities or their HOA to emphasize the need for reasonable accommodation under PRB-1.

Remote Operations

The use of remote control in amateur radio is governed by a combination of regulations, which depend on the transmitter's location. FCC Part 97 regulations apply in full for transmitters within the United States. A valid FCC-issued amateur radio license is mandatory for the control operator.

When the transmitter is located outside the US, local government regulations apply. Reciprocal agreements may exist with the US that dictate privileges for US-licensed operators. Always do your due diligence to research the specific regulations of the transmitter's location.

Restrictions on third-party traffic remain in effect regardless of remote control use. These limitations depend on the operator's license class and the countries involved. The control operator must ensure compliance with all applicable regulations at all times.

Rights and Restrictions of Control Operators

A control operator has extra on-air privileges, but the job also comes with specific responsibilities. This section covers the privileges and restrictions that must be strictly observed in compliance with FCC rules and ethical standards.

Frequency

General Class licensees enjoy access to a broader range of HF frequencies, but it is imperative that they are aware of the frequency allocations. There is no need to memorize every band segment. Most General Class operators use a frequency chart as a reference, and a downloadable chart is available from the ARRL website. When you operate on the bands, always consult the chart to confirm transmission only within the designated segment for your license class.

Amateur radio operators with a General Class license have operating privileges at the upper portion of the permitted band segments for each mode. General Class licensees do not have full amateur privileges on the 60-meter band, as operation is limited to specific channels and modes. Repeater operation within the HF spectrum is restricted to the 10-meter band within the 29.6 to 29.7 MHz range.

There are specific regulations that govern two HF bands. The 60-meter band requires channelized operation and is limited to USB, CW, and designated digital modes. The maximum effective radiated power (ERP) allowed is 100 W. On the 30-meter band, permissible transmissions include CW, RTTY, and data signals, which adhere to a 200-watt peak envelope power (PEP) limit.

Frequency-Wavelength Interconversion and Examples

The formulae to convert between wavelength and frequency in amateur radio allocations are:

- 300 ÷ Wavelength (Meters) = Frequency (MHz).
- 300 ÷ Frequency (MHz) = Wavelength (Meters).

These formulas are essential to understand amateur radio band allocations, antenna design, and propagation characteristics. Consider the examples below.

Example 1: What is the frequency of a radio wave with a wavelength of 20 meters?

- Plug the wavelength (20 meters) into the formula: 300 ÷ 20 meters = 15 MHz.

- Therefore, a radio wave with a wavelength of 20 meters has a frequency of 15 MHz.

Example 2: What is the wavelength of a radio wave with a frequency of 7.150 MHz?

- Plug the frequency (7.150 MHz) into the formula: 300 ÷ 7.150 MHz ≈ 41.96 meters.

- Therefore, a radio wave with a frequency of 7.150 MHz has a wavelength of approximately 41.96 meters.

The constant "300" used in frequency-wavelength conversions is a simplified approximation of the speed of light (299,792,458 m/s) in millions of meters per second. This approximation makes calculations easier for amateur radio purposes. Frequency and wavelength have an inverse relationship. That is, as one increases, the other decreases. When you apply these formulas, remember to use consistent units (meters for wavelength and megahertz for frequency).

General Class Frequency Privileges

As a General Class licensee, you gain full operating privileges on six HF bands, as allocated by the World Administrative Radio Conference (WARC). These include 160 meters (1.800–2.000 MHz), a classic band for long-distance communication, especially at night. Operating privileges also exist on 60 meters (5.3305–5.4065 MHz), 30 meters (10.100–10.150 MHz), 17 meters (18.068–18.168 MHz), 12 meters (24.890–24.990 MHz), and 10 meters (28.000–29.700 MHz). The 30-meter band is limited to CW, RTTY, and data modes. The table below illustrates this information better.

Band (Meters)	Frequency Range (MHz)	Modes
160	1.800–2.000.	All modes.
60	5.3305–5.4065.	All modes.
30	10.100–10.150.	CW, RTTY, Data modes.
17	18.068–18.168.	All modes.
12	24.890–24.990.	All modes.
10	28.000–29.700.	All modes.

General Class ham operators cannot operate on "old-timer" bands. These bands, accessible only by Extra Class license holders, are listed below:

- 80 meters (3.500–4.000 MHz).
- 40 meters (7.000–7.300 MHz).
- 20 meters (14.000–14.350 MHz).
- 15 meters (21.000–21.450 MHz).

The restriction of certain bands for General Class licensees serves multiple purposes. It encourages operators to advance their knowledge and upgrade to Amateur Extra, helps manage congestion on popular bands, and reflects historical allocation decisions. Also, some bands demand higher technical expertise, which is not fully covered in the General Class exam. These limitations promote growth within the hobby and ensure efficient and responsible use of the radio spectrum.

Bands

Amateur radio operators use various HF bands for different communication modes. For instance, SSTV is used for image transmission, mainly on bands like 75/80, 40, 20, 15, and 10 meters. Image transmission is not allowed on 30 meters due to regulatory constraints. Certain bands, like 60 meters, operate on a channelized basis. Others, like the 80/75-meter band, have different privileges based on license class. Popular bands like 20 meters are favored for long-distance communication, and some, like 10 meters, even support FM repeaters.

Image Bands

Slow Scan Television (SSTV) is used in amateur radio to transmit still images over radio frequencies. This mode converts image information into audio tones that represent the brightness and color information of each pixel within the image. Hence, SSTV is considered a voice mode. Transmission of a complete image can take several seconds or even minutes as the encoding process is line-by-line. Thus, the name "slow scan." An SSTV receiver decodes the audio tones to display the image.

SSTV activity occurs within specific segments of the HF (shortwave) bands and allows image exchanges across long distances. The most common SSTV frequencies are 3.845 MHz (75/80 meters), 7.171 MHz (40 meters), 14.230 MHz (20 meters), 21.340 MHz (15 meters), and 28.680 MHz (10 meters). Local SSTV transmissions might also occur on VHF and UHF bands. Consult with local clubs or repeater groups to find designated frequencies.

Ham operators cannot transmit SSTV on the 30-meter band. FCC regulations for this band (10.100–10.150 MHz) limit transmissions to CW, RTTY, and data modes. SSTV, which is considered a data mode, is not permitted on the 30-meter band.

SSTV has many modes, each with distinct characters when it comes to image quality and transmission speed. Popular modes include Martin (M1 and M2), Scottie (S1 and S2), Robot (36 and 72), and PD (50, 90, 120).

You will need an HF transceiver (for HF bands) or a VHF/UHF transceiver (for local activity) with SSB capability to participate in SSTV. Also, you will need a computer with a sound card and SSTV software that manages the image-to-audio conversion and interfaces with the transceiver.

60 Meter

Some amateur radio bands are limited to a channel-based operation, similar to how CB radios operate. This limitation confines transmissions to specific channels, as against the flexible tuning option of a variable frequency oscillator (VFO), which is allowed on most amateur bands. The 60-meter band (5.3305–5.4065 MHz) is an example of a channelized operation band in amateur radio. Operators must transmit on specific frequency channels and may not tune freely.

A maximum power limit of 100 W PEP, measured as ERP, applies on the 60-meter band. If you use an antenna with gain, you must reduce your transmitter power as needed to stay within the ERP limit. Transmissions on this band are also restricted to a maximum bandwidth of 2.8 kHz. The 60-meter band allows USB (upper sideband) voice transmissions, CW (Morse code), and specific digital modes like RTTY and PSK31.

80/75 Meters

The 80/75-meter band covers a 500 kHz range of the HF spectrum (3.5 MHz to 4.0 MHz). It is often referred to as a single band, though technically, it includes both the lower 80-meter and the 75-meter designations.

An operator's license class dictates operating privileges on this band. General Class licensees have voice (phone) privileges from 3.8 to 4.0 MHz and CW privileges from 3.525 to 3.750 MHz. Extra Class operators gain access to the entire band for both voice and CW.

Note that desirable DX stations often operate in the lower portion of the band. Refer to the frequency chart to make sure you only transmit within the segments permitted for your license class.

20 Meter

The 20-meter band (14.000–14.350 MHz) is a popular HF band for long-distance (DX) communication. General Class licensees can access designated portions of the band for voice (phone) and CW operation.

10 Meter

The 10-meter band (28.000–29.700 MHz) supports FM repeaters above 29.5 MHz. These repeaters function similarly to those on VHF/UHF bands and extend the local communication range. They use a piece of specialized filter called a duplexer.

Shared Bands

Amateur radio operators often hold secondary allocations within certain bands. Primary service stations are given priority in these shared bands. Amateurs must avoid harmful interference and QSY (stop transmission) immediately if a primary service station becomes active. Even if the signal types appear similar, a ham may not initiate contact with stations in the primary service. One example is the 60-meter band, where amateur radio holds a secondary allocation. Another example is the 30-meter band, limited to CW, RTTY, and data modes.

Beacon Stations

Beacon stations are automated transmitters placed strategically around the world. They help amateur radio operators assess propagation conditions with brief Morse code identification messages, such as their call sign and location. When you monitor beacons on designated frequencies, this allows you to gauge if a band is open to a particular region.

Specialized clocks and smartphone apps help track beacon schedules. The Northern California DX Foundation (NCDXF) maintains a network of beacon stations. Their website and other propagation resources provide current frequency information.

Amateur radio operators can establish beacon stations on the 10-meter band (28.2–28.3 MHz) without special authorization. Transmissions must be automatically controlled and include only a call sign, location, and the designated power limit. The maximum amount of power a beacon station can radiate is 100 watts PEP.

Power Levels

You must be able to distinguish between PEP and ERP. PEP measures the maximum power output of a transmitter at the peak of a single RF cycle and is the standard measurement for most amateur transmissions. Beacon stations have a maximum limit of 100 watts PEP. Broadcast licenses use ERP, a more complete representation of signal strength that combines transmitter power, antenna gain, and system losses. The 60-meter band is an example where ERP restrictions apply within amateur radio.

Band	Maximum Power Limit	Maximum Symbol Rate	Maximum Bandwidth
10 meters.	1500 watts PEP output.	1200 bauds.	2.8 kHz.
12 meters.	1500 watts PEP output.	-	2.8 kHz.
20 meters.	1500 watts PEP output.	-	2.8 kHz.
60 meters.	ERP of 100 watts PEP for a dipole antenna.	-	2.8 kHz.
28 MHz.	1500 watts PEP output.	-	2.8 kHz.
1.8 MHz.	1500 watts PEP output.	-	2.8 kHz.
Below 28 MHz (RTTY/Data).	-	300 bauds.	-

The table above showcases a selection of power, bandwidth, and symbol rate limits for certain amateur radio bands and modes for General Class licensees. Other bands and license classes may have varied power restrictions. Always consult the most recent FCC regulations for power limits and other operational rules pertinent to amateur radio operations.

Third-Party Traffic

Third-party traffic involves messages for someone without an amateur radio license. The FCC allows third-party traffic under certain conditions. You can exchange messages with stations in countries that have a third-party traffic agreement with the US. During emergencies, you may handle third-party traffic directly related to the immediate safety of lives and property protection.

A third-party message may include amateur radio discussion or information related to emergencies and disaster relief. These messages should not be about business and should never include obscene, indecent, or profane language. As the licensed operator, you are responsible for all transmissions and must correctly identify your call sign and the third party's station.

Repeaters

Repeaters are special amateur radio stations that receive a signal on one frequency and retransmit it on another, which greatly extends the range of handheld and mobile radios. Most repeaters are coordinated to manage spectrum use and minimize interference. If an uncoordinated repeater disrupts a coordinated one, the uncoordinated station must resolve the issue.

Amateur radio operators often hold secondary status on certain bands. This status requires them to avoid harmful interference to primary service users. Cross-band repeaters receive on one band and retransmit on another. A technician can use a cross-band repeater if the control operator holds an Extra Class license or higher.

Special care is needed to avoid harmful interference near FCC Monitoring Stations, on bands where amateur radio has a secondary status, or during Spread Spectrum transmissions.

Spread Spectrum

Spread Spectrum (SS) is a transmission technique where the signal energy is spread across a wider portion of the spectrum than traditional modes. This results in lower power density at any specific frequency. The maximum allowable power radiation limit for spread spectrum transmissions is 10 watts PEP.

This provides advantages like increased resistance to interference, less noticeable transmissions, and suitability for digital data transmission. Spread Spectrum is useful for digital data modes, Automatic Packet Reporting Systems (APRS), and helps decode weak signals.

Two primary types of Spread Spectrum exist—Direct-Sequence Spread Spectrum (DSSS) and Frequency-Hopping Spread Spectrum (FHSS). DSSS uses a pseudo-random code to spread the signal. In FHSS, the transmission frequency rapidly changes according to a pattern.

Unlicensed Output Power

Regulations for maximum output power for unlicensed Wi-Fi devices differ between countries. Europe and Germany, for instance, limit output to 100 milliwatts, while the United States allows up to 1 watt. Although amateur radio operators have the technical capability to modify Wi-Fi devices for higher power transmissions (up to 10 watts), these devices should not be used for general internet traffic.

Amateur Radio Use of 2.4 GHz

Amateur radio operators have a secondary allocation on the 2.4 GHz band and must avoid interference with the primary unlicensed Wi-Fi devices. Direct communication between unlicensed Wi-Fi devices and amateur stations is not permitted. However, amateurs can create private mesh networks on this band. Mesh networks enable communication between licensed stations within a defined area.

Banned Countries

Currently, the US does not have blanket prohibitions on contact with amateur radio operators in any specific country. However, temporary restrictions may apply to certain countries, notably North Korea and Yemen. These restrictions can fluctuate due to political embargoes or other factors. Amateur operators must stay current on any temporary bans.

Technical Standards for Amateur Radio

Technical standards for amateur radio encompass power limits, measurement methods, and specific regulations for digital communications to ensure efficient spectrum use and minimize interference.

Maximum Output Power

Different bands have prescribed limits for output power based on the license class. For instance, General, Advanced, and Extra Class licensees generally have a maximum PEP of 1500 watts on HF bands, with exceptions on the 30-meter band (200-watt PEP limit) and the 60-meter band (100-watt ERP limit relative to a half-wave dipole, and 2.8 kHz bandwidth restriction).

Similarly, Technician licensees are limited to 200 watts PEP on HF. However, within former Novice segments on 80, 40, and 15 meters, only Extra Class licensees can use the full 1500-watt PEP.

Q-signals are often used to describe power levels. QRP indicates low power, usually 5 watts or less, while QRO indicates higher power. The FCC limits amateur output power to 10 watts for SS transmissions on higher UHF and microwave bands.

Many HF stations use linear amplifiers to give their signal a power boost. To stay within the rules, measure your output power right where it leaves the transmitter or the amplifier—whichever one is last in line before your signal heads out to the antenna. Prioritize RF safety, especially on microwave frequencies where high-gain antennas can create exposure risks.

Peak Envelope Power (PEP)

PEP measures the maximum power your transmitter outputs during a single peak of an RF cycle. This measurement is relevant to amplitude-modulated (AM) signals where power levels vary throughout the signal.

How to Calculate PEP

- **Measure Peak-to-Peak Voltage:** Use an oscilloscope to measure the peak-to-peak voltage (Vp-p). The Vp-p is the difference between the wave's highest positive and lowest negative points.
- **Divide by Two:** Since you want the peak voltage from the "zero" line rather than the full peak-to-peak, divide your Vp-p measurement by 2.
- **Convert to RMS:** Multiply the result by 0.707 to get the RMS (root mean square) voltage. The RMS is the equivalent average voltage of the signal.
- **The Formula:** Use this formula to calculate PEP: $PEP = (RMS\ Voltage)^2 \div Resistance\ (Ohms)$. Since most amateur radio equipment has a standard impedance of 50 ohms, the formula becomes $PEP = (RMS\ Voltage)^2 \div 50$.

Example: If your peak-to-peak voltage measures 120 volts, what is the RMS voltage?

The peak-to-peak voltage is given to be 120 volts.

- To find the peak voltage, divide the peak-to-peak voltage by 2: 120 volts ÷ 2 = 60 volts.
- To convert the peak voltage to RMS voltage, multiply by 0.707: 60 volts × 0.707 ≈ 42.42 volts.
- To calculate the PEP, use the formula $PEP = (42.42)^2 \div 50 \approx 36$ watts.

Digital Communications

Digital transmissions encompass various modes like radio teletype (RTTY) and PSK (phase-shift keying) and have become a prominent aspect of amateur radio communication. Hence, the FCC has set technical standards for these transmissions to promote good spectrum use and reduce user interference. A key focus of these regulations is the bandwidth occupied by digital signals. Bandwidth is directly tied to symbol rate, which measures the number of signal changes per second. Faster symbol rates require wider bandwidths to carry the same amount of information.

The FCC dictates limitations on the symbol rate so digital signals don't consume excessive bandwidth and cause interference. These restrictions are outlined in FCC Part 97, sections 97.305(c) and 97.307(f). Due to their limited spectrum, HF bands have stricter symbol rate limits. As you move up in frequency to VHF, UHF, and beyond, the FCC allows progressively faster symbol rates due to the increase in available bandwidth. At 902 MHz and above, bandwidth limitations are relaxed to enable experimentation with high-speed digital modes.

The table below lists important symbol rate limitations for digital transmission.

Frequency Band	Maximum Symbol Rate	Additional Notes
Below 28 MHz.	300 baud.	Applies to RTTY and data emissions.
10-meter band (28 MHz).	1200 baud.	Applies to RTTY and data emissions.
Above 50 MHz.	No limits.	Generally no limit, but some bands may have specific restrictions. Check Part 97.

The FCC acknowledges the continuous development of new digital protocols by amateur radio operators. It mandates that the technical specifications of any new digital protocol intended for on-air use must be publicly documented before its implementation. This documentation can take various forms, such as publication on a website dedicated to amateur radio or inclusion in a relevant technical journal.

Chapter 2: Operating Procedures

In this chapter, you will learn techniques to navigate the often-congested HF spectrum and principles to share frequencies with respect, as well as how to contribute during emergencies.

Basic Operating

Unlike VHF/UHF, which mainly use line-of-sight, HF signals propagate long distances through refraction off the ionosphere. The specific band, time of day, and solar conditions significantly influence this.

As a General Class licensee, you can contact stations worldwide on the HF bands. Two common methods to begin communication are "CQ" and "CQ DX."

- **CQ (pronounced "seek you")** is a universal call used to indicate your availability for a contact. It signifies that you would like communication and want to establish a two-way conversation with any station that hears you.
- **CQ DX (pronounced "seek you dee-ex")** is a variation used by General Class licensees and above. It expresses your desire to connect with stations located in a different country from your own. This helps to focus your call on stations beyond your local area and potentially achieve long-distance communication (DX-ing). CQ DX means a call to stations outside your own country.

Good Practices

Choose a Frequency to Use

- **Listen First**: Before you transmit, tune across the band to assess occupancy. Select a clear frequency where you can politely join a conversation.
- **Always Ask**: To minimize the potential for harmful interference before you start a CQ call, inquire if the frequency is occupied. Although the frequency might sound clear, you might not hear a conversation because of propagation conditions. Clarify with questions like "Is the frequency in use? This is [your call sign]." Repeat the inquiry, and include your call sign. If you use CW or digital modes, transmit "QRL? DE [your call sign]."
- **Respect Band Plans**: Familiarize yourself with generally accepted volunteer band plans for voice, CW, and digital modes. Operate within these guidelines to minimize the risk of accidental interference.

- **No Monopoly**: Except in designated emergencies, no amateur station possesses exclusive rights to any frequency. Avoid any perception of "frequency ownership," and consistently maintain a flexible approach with "Plan B" frequencies.
- **Adhere to FCC Regulations**: Use frequencies allowed by FCC regulations part 97. The table below shows the available frequency privileges for Generals.

Band (MHz)	Allocation Type	Modes	Power Limit (Watts)
160 meters (1.8–2.0).	Shared – Varies by region.	CW, RTTY/Data.	Refer to FCC Part 97 for specifics.
80 meters (3.5–4.0).	Shared.	CW, SSB, RTTY/Data.	1500.
60 meters (5.330–5.405).	Limited to 5 channels.	CW, SSB, Data.	100.
40 meters (7.0–7.3).	Shared.	CW, SSB, RTTY/Data.	1500.
30 meters (10.1–10.15).	Shared.	CW, RTTY/Data.	1500.
20 meters (14.0–14.35).	Shared.	CW, SSB, RTTY/Data.	1500.
17 meters (18.068–18.168).	Shared.	CW, SSB, RTTY/Data.	1500.

15 meters (21.0–21.45).	Shared.	CW, SSB, RTTY/Data.	1500.
12 meters (24.89–24.99).	Shared.	CW, SSB, RTTY/Data.	1500.
10 meters (28.0–29.7).	Shared.	CW, SSB, RTTY/Data, FM.	1500.

- **The 50.1–50.125 MHz CW DX Window**: This narrow segment at the bottom of the six-meter CW allocation aims to provide a less crowded area for long-distance (DX) contacts through the use of Morse code. DX stations often have fainter signals due to distance and propagation factors. When they are concentrated in one area, this helps them stand out. A focus on non-lower 48 contacts means stations in Alaska, Hawaii, US territories, and other countries are prioritized—the true DX.

- **Signal Separation**: Although the ideal spacing varies based on conditions, always aim to leave enough space between your signal and neighbor transmissions to minimize interference. This is especially important for the often-crowded HF bands. CW uses the narrowest bandwidth, while SSB voice requires wider separation. The table below shows recommended signal separation for common modes.

Mode	Recommended Separation	Notes
CW.	150 Hz–500 Hz.	Can often be narrower with good conditions.
SSB.	2 kHz–3 kHz.	Maintains voice intelligibility without excessive spread.
RTTY.	200 Hz–500 Hz.	Varies slightly by specific RTTY sub-mode.
PSK31.	31 Hz (fixed).	Designed for minimal bandwidth.
Other Digital Modes.	Consult mode specifications.	Separation can vary widely.

Manage Interference

Interference may stem from man-made sources like other stations (QRM) or natural phenomena like noise, static, and electrical storms (QRN). Harmful interference is any severe disruption of communication. Deliberate interference, though rare, should be avoided entirely. To effectively manage interference, understand band characteristics, optimize your station's setup (including the use of clean signals and directional antennas), and maintain a flexible approach. If interference persists, consider a change of frequencies since no one has exclusive rights to any specific spot on the dial.

Propagation Problems

Propagation refers to how radio waves travel through the atmosphere and over the earth's surface. It constantly changes, influenced by:

- **Time of Day**: The sun's ionization of the upper atmosphere impacts which HF bands work best for DX communication at different times.
- **Solar Cycle**: The 11-year solar cycle brings high and low sunspot activity periods, which can drastically change HF propagation conditions.
- **Seasons**: Variations in daylight hours and atmospheric conditions over the seasons affect optimal bands.
- **Sporadic Events**: Unpredictable phenomena like Sporadic-E can suddenly open up unexpected communication paths, even on VHF/UHF.

Propagation determines how strong (or weak) a signal from a distant station will be on your chosen frequency. Certain frequencies might only support DX contacts during specific times of the day or under favorable solar conditions. As propagation shifts, stations you did not hear previously may suddenly appear on or near your frequency. In such situations, always try to resolve the interference issue through polite and open discussion with the other stations.

Net Management

Nets are scheduled gatherings on specific frequencies where ham radio operators with shared interests come together. They provide structure and a sense of community within the hobby. Nets often form around specific modes, technical interests, geographic regions, or activities. Examples include maritime nets, vintage radio nets, contest nets, youth nets, and traffic nets.

Nets can be found in the ARRL Net Directory, through local clubs, or within online forums or groups, and they have procedures. Listen carefully and follow the instructions of the Net Control Station (NCS) to participate appropriately. As part of good net management, always designate a backup frequency to ensure smooth operation in case of interference or unexpected band conditions.

Band Plans

Band plans are informal agreements created by the ham radio community that suggest where to find different activities within each band. Band plans are not legal documents. Nevertheless, hams who follow them demonstrate good amateur practice and respect for fellow operators. The FCC allocates specific portions of bands for certain purposes (for example, CW, digital modes, and DX Window), which are mandatory.

Contests

Contests are popular on the HF bands, where hams compete to make contacts, often under specific rules or objectives. Although they are fun activities, the stations and contacts must still be done per standard FCC regulations. Examples include the CQ World Wide DX Contest, ARRL Sweepstakes, and Field Day.

Make Contacts

Break into a Conversation

In voice modes such as SSB and FM, "breaking into (joining) a conversation" dictates how you respectfully join an exchange between two or more stations. To proceed, monitor the conversation for a natural pause. After this pause, transmit your call sign once. For example, say you hear a discussion on 20 meters SSB. You should wait for a pause and say, "This is [your call sign]. Is there any chance I can join in?" Should a station acknowledge your transmission, you will likely receive an invitation to participate.

Break-In Keying on CW

Break-in keying is a technical capability specific to the Morse code (CW) operation. This functionality facilitates a more conversational and rapid exchange of CW communications. Break-ins can be full or semi.

Full break-in (or QSK) enables you to hear the other station's signal in real time, even between individual dots and dashes. It requires a QSK-capable radio and is ideal for fast-paced CW conversations or contests, where split-second timing is vital. You can instantly hear if the other station attempts to transmit, even if they begin mid-word. This allows for smoother corrections or rapid responses to urgent situations.

With semi-break-in, there is a slight delay (VOX delay) when you switch from transmit to receive mode. You can generally hear the other station between words or short phrases but not necessarily between individual dots and dashes. The slight delay can minimize audible keying clicks and potentially reduce distractions.

Although break-in capabilities enhance efficiency, hams must adhere to proper etiquette. They should use appropriate pauses before they transmit or briefly transmit "BK" to indicate an intention to interject. Also, note that not every transceiver possesses full break-in keying (QSK) functionality. For instance, say you are in a CW QSO and realize the other station has made an error. You could send 'BK' to politely let them know you want to interject.

Common Contact Exchanges

There is a usual flow of information when a CQ call is made on a ham radio. To start, exchange essential information: your call sign, location, and a signal report. Signal reports are often exchanged early in an HF contact to help both stations adjust power and expectations based on current propagation conditions. After this core exchange, conversations can become more informal and cover topics like weather, equipment details, hobbies within the amateur radio world, or personal interests.

Station Logs

Station logs are records of your on-air contacts. Information you might log includes the date, time, frequency, call sign of the other station, signal report, and additional notes like mode or location. Although not legally required for most operations, a detailed station log can be valuable if the FCC ever requests information about your operation.

Station logs come in paper and electronic forms. Traditional paper logs provide simplicity. Electronic logging solutions (both free and paid) often provide advanced features, such as those discussed below.

- **Radio Integration (CAT)**: Logging software can automatically pull frequency and mode data directly from your transceiver.
- **Online Links**: This involves integration with call sign databases (like LoTW and QRZ) and spotting networks.
- **Customization**: You can tailor the information logged to suit your interests. Popular software choices include N3FJP's Amateur Contact Log and the N1MM contest logger.

Modes

CW Mode (Morse Code)

CW, or Morse code, is a classic mode of communication found mainly in designated segments at the lower end of each HF band. CW transmissions can be generated with various devices, such as traditional hand keys, semi-automatic "bugs," electronic keyer paddles, and computer keyboards. Although Morse is technically permissible on most amateur frequencies, it is considered good practice to respect these traditional allocations. Operators communicate with a series of dots and dashes that represent letters and numbers.

CW transmissions can be encountered at a wide range of speeds, from approximately five words per minute (WPM) to upward of 50 WPM. When you respond to a CQ in CW, it is essential to match the send speed of the calling station. If you find the pace too rapid, a polite "QRS" (please send slower) is customary to request a reduced speed.

AM and SSB Voice Modes

On the HF bands, SSB is the predominant mode of voice communication. SSB has advantages over AM. Its narrower bandwidth (approximately 3 kHz compared to AM's 6 kHz) results from the suppression of the carrier and one sideband. This reduces bandwidth and increases power efficiency.

To maintain efficiency and consistency, amateur radio operators adhere to customary practices for sideband selection. The magic number here is nine. USB is used on frequencies above 9 MHz (including the 20-meter to 10-meter bands). The 60-meter band, however, also uses USB, despite being below 9 MHz. USB is also used on VHF and UHF bands. LSB is the standard for frequencies below 9 MHz (such as 160-meter, 80-meter, and 40-meter bands).

Keep your band chart posted somewhere visible. Test questions on side bands may refer to bands by their wavelength or frequency range. A helpful trick is to convert from the given unit to the unit you are familiar with via one of the formulae below:

- Frequency (MHz) = 300 ÷ Band (Meters).
- Band (Meters) = 300 ÷ Frequency (MHz).

Split Mode

Contemporary transceivers often feature dual VFOs and provide operational flexibility. A primary advantage is the ability to operate in "split" mode. This allows the operator to simultaneously transmit on one frequency while a slightly different frequency is monitored for responses. This capability is helpful to manage pileups or stations with weaker signals.

Split mode involves the configuration of the transceiver with distinct transmit and receive frequencies. DX stations frequently leverage this function to spread out station responses during pileups. This improves their ability to distinguish and work with individual callers. Transceivers equipped with dual VFOs greatly simplify operating in split mode.

Image Modes (TV)

Image modes allow SSTV or real-time video (FSTV) transmission over amateur radio frequencies. SSTV is a popular choice on the HF bands and is permitted wherever phone (voice) transmissions are allowed, except the 30-meter and 60-meter bands.

FSTV is mainly used on UHF and higher frequency bands. FSTV transmits full-motion video in a manner analogous to broadcast television. It provides real-time video transmission and requires more specialized equipment.

Unique Characteristics of the 60-Meter Band

The channelization of the 60-meter band (5 MHz) differentiates it from other HF allocations. This means that operators are restricted to transmissions on five specific, narrow frequencies. The FCC also sets a maximum ERP of 100 watts on the 60-meter band. ERP considers both your transmitter power and any gain your antenna might add.

The focus on a dipole antenna as the reference point ensures fairness. Every station is effectively limited to the same signal strength. This prevents those with high-gain antennas from an excessive advantage. If you use anything other than a simple dipole antenna on 60 meters, you must meticulously record your antenna gains to demonstrate compliance with FCC regulations.

HF Receiving

QSL Cards

QSL cards are postcards exchanged between ham radio operators to confirm a contact (communication). They contain essential information about the contact, such as call signs, date, time, frequency, and mode. They can be physical or electronic. QSL cards are especially good to confirm DX contacts with stations in faraway countries. Many contests or ham radio award programs require them as proof of your contacts.

Q Signals

Q signals (or Q-codes) are three-letter codes that start with the letter "Q." They were originally developed for commercial and maritime radiotelegraph communication. Ham radio adopted many of these codes as a shorthand way to convey standard questions, information, or instructions quickly and efficiently. Not all of them are universal. Some Q signals have specific meanings within certain contexts (like contests). Examples of frequently encountered Q signals are discussed in the following table.

Q Signal	Plain Language Meaning	Notes
QRL	"Are you busy?" or "Is this frequency in use?"	A way to check if someone is already engaged in a conversation or if the frequency is clear before transmitting.
QRM	"I have interference from other stations."	Lets the other operator know it is difficult to hear them due to other ham radio transmissions.
QRN	"I have interference from static or natural sources."	Indicates disruption of reception by atmospheric noise, electrical storms, or other natural phenomena.
QRP	"Reduce power."	Often used to signify operations with low power output (usually 5 watts or less).
QRV	"I am ready to copy."	Conveys that you and your station are prepared to start a contact.
QSL	"I acknowledge receipt of your transmission."	Confirms that you have received and understood the other station's message.
QSY	"Change frequency to ..."	Used to ask that the conversation be moved to a different frequency, often to escape interference or find a clearer spot on the band.

QTH	"What is your location?"	A request for the other station's geographical location (city, state, and grid square).

HF Receivers Versus VHF/UHF FM Radios

VHF/UHF FM transceivers, designed for channelized communications, usually feature basic controls focused on channel selection, volume, and squelch (mute background noise). In contrast, HF transceivers designed for SSB and CW operation provide a wider array of controls due to the continuous tuning nature of HF bands. HF receivers provide continuous tuning across bands for precise frequency selection rather than being restricted to fixed channels. Due to the characteristically higher noise floor on HF, squelch controls are often absent. Furthermore, modern HF receivers frequently incorporate digital signal processing (DSP), which enables advanced filters, noise reduction, and signal enhancement capabilities to help extract desired signals from difficult environments.

HF Transmissions

PTT (Push-to-Talk)

HF transceivers provide methods for voice transmission initiation that parallel those found in VHF/UHF equipment. The classic "push-to-talk" (PTT) is a standard mode where the operator physically activates the transmitter through the press of a button.

VOX (Voice-Operated Transmit)

Many HF transceivers feature voice-operated transmit (VOX). This functionality enables hands-free operation, as the transmitter automatically activates when someone speaks into the microphone. VOX is advantageous during contests or extended roundtable discussions where frequent transitions between transmit and receive are necessary. The term 'VOX' generally applies to voice transmissions but can also be used with CW and various digital modes.

VOX enables hands-free operation on voice modes, and HF transceivers usually include adjustable VOX parameters to fine-tune the system's response, as discussed below.

- **VOX Gain**: This parameter determines the audio level required to activate the transmitter. Higher-gain settings require less audio input to activate the transmitter, while lower settings necessitate a stronger signal from your microphone.

- **VOX Delay**: This introduces a brief hold time after talk stops before the transmitter switches back to receive mode. This helps so the system doesn't switch on and off rapidly due to short pauses in speech.

- **Anti-VOX**: So the VOX circuit isn't inadvertently triggered, this function is used. It is essential for comfortable operation, especially when someone uses headphones.

CW (Morse Code) Transmitting

Morse code operators use various methods to send, such as straight keys, semi-automatic "bugs," iambic keyer paddles, and computer keyboards.

- **Straight keys** are classic hand-operated keys.

- **Bugs** are semiautomatic keys for faster, more consistent code generation.

- **Iambic paddles** are electronic keyers that provide rhythmic dot-dash generation.

- **Keyboards** are used in computer-based CW transmission.

Modern transceivers have full break-in (QSK) for real-time conversational flow in CW. Morse code operations require some special practices and techniques, as discussed below.

- **Zero Beating**: In CW operation, zero beating refers to the meticulous adjustment of your transmit frequency to match the station you intend to contact. A zero beat produces a pure, clear tone in the receiver. This minimizes interference from nearby stations and enhances the readability of the desired signal.

- **RST Reports**: RST reports provide a standardized method to convey the quality of a received CW signal. The acronym stands for Readability (how well you can understand the signal), Strength (how loud it is), and Tone (the quality of the CW note's pitch). Each element is rated numerically, and higher numbers mean better quality. Readability runs on a scale of 1 (unreadable) to 5 (perfectly readable). Strength runs on a scale of 1 (extremely weak) to 9 (extremely strong), while tone runs on a scale of 1 (very rough) to 9 (perfect tone). A "C" suffix may be added to the report to signify a chirpy or unstable signal. For example, if you give a station a report of "579," their signal is perfectly readable, fairly strong, and has a clear tone.

- **Electronic Keyers**: These devices automate the generation of dots and dashes for CW transmission. Most modern transceivers incorporate internal electronic keyers that provide the basic functionality. External keyers allow greater flexibility, customization options, and advanced features. Various input devices, such as iambic paddles (for precise timing control) or computer keyboards, can control these keyers, where software translates something typed into transmitted CW.

Some specific procedures also apply to CW transmission, as discussed below.

- **Call Response**: When you answer a CQ call, it is proper etiquette to match the speed at which the CQ was sent. This demonstrates respect for the call operator and ensures smooth communication.

- **Procedural Signals**: Procedural signals, or prosigns, are two-letter combinations transmitted as a single Morse code character. They control the flow of communication, streamline exchanges, and avoid the transmission of full words or phrases. Some examples of prosigns and their usage are illustrated in the following table.

Prosign	Meaning	Usage Example
AR.	End of message.	This indicates the conclusion of a formal message within a longer transmission. Example: “R R HR NAME IS JOHN QTH BOSTON MA AR” (Acknowledged, here, name is John, location Boston Massachusetts, end of message).
BT.	Break.	This separates different parts of a message, like sender information, content, and signature. Example: “JOHN SMITH BT 123 MAIN ST BT ANYTOWN CA BT” (John Smith, break, 123 Main St, break, Anytown California, break).
K.	Over.	This indicates you have finished transmission and need a response from the other station. Example: “CQ CQ CQ DE W1XYZ W1XYZ K” (General call, general call, general call, from W1XYZ, W1XYZ, over).
KN.	Only.	This is used to specify that you need a particular station and want others to remain silent. Example: “W1XYZ DE W2ABC KN” (W1XYZ, from

		W2ABC, only).
SK.	End of contact.	This is used to mark the conclusion of the entire CW exchange. Example: "73 ES CUL DE W5XYZ SK" (Best regards and see you later, from W5XYZ, end of contact).

- **Informal Abbreviations**: Like prosigns and Q signals, informal abbreviations expedite exchanges and save precious airtime. These abbreviations contain short letter combinations that represent commonly used phrases or concepts. Examples are listed in the table below.

Abbreviation	Meaning	Usage Example
CL.	Clear and going off the air.	This signals an end to the communication session as someone goes offline. Example: "73 ES TNX FOR THE QSO CL" (Best regards, thanks for the conversation, clear).
CUL.	See you later.	This is a casual way to say goodbye at the end of a CW exchange. Example: "FB QSO OM CUL DE W1ABC" (Fine conversation, old man, see you later from W1ABC).
TNX.	Thanks.	This expresses gratitude to the other station. Example: "TNX FOR THE RPRT ES FB SIGNAL SK" (Thanks for the report and good signal, end of contact).
WX.	Weather.	This is used to inquire about or provide weather information. Example: "WX HR IS SNY TNX BT" (Weather here is sunny, thanks, break).

73.	Best wishes.	This is a formal way to conclude a CW conversation. Example: "CUL ES 73 DE K1XYZ" (See you later and best wishes from K1XYZ).

Procedures and Abbreviations

Although the fundamental concepts of contact initiation and exchange mirror those of VHF/UHF operation, the HF environment necessitates certain adaptations. Due to potential signal degradation on HF, clear phonetics (such as the NATO Phonetic Alphabet) are paramount for accurate communication.

Furthermore, Q-signals are frequently used in SSB conversations even though they were originally designed for CW operation. Knowledge of common Q-signals will enhance your comprehension of on-air exchanges. However, hams are strongly advised to refrain from the use of "CB lingo" and 10-codes on amateur radio bands. These practices are considered outdated, less efficient, and can hinder clarity. Plain language is the preferred and most effective mode of communication within the ham radio community.

NATO Phonetic Alphabets

Static, accents, and unusual call signs can muddle transmissions. The NATO Phonetic Alphabet provides a standardized system for spelling words with easily recognizable phonetic words. The alphabet is listed in the table below.

A: Alfa.	N: November.
B: Bravo.	O: Oscar.
C: Charlie.	P: Papa.
D: Delta.	Q: Quebec.
E: Echo.	R: Romeo.
F: Foxtrot.	S: Sierra.
G: Golf.	T: Tango.
H: Hotel.	U: Uniform.
I: India.	V: Victor.
J: Juliet.	W: Whiskey.
K: Kilo.	X: X-ray.
L: Lima.	Y: Yankee.
M: Mike.	Z: Zulu.

Emergency Communications

Amateur Radio in Emergency Situations

The FCC recognizes amateur radio operators' vital role in essential communications during emergencies or disasters. FCC rules stipulate that amateur stations may engage in transmissions that directly support the management of such events and provide needed relief.

Priority Access for True Emergencies

No user or station has priority access to amateur radio frequencies during normal operations. All licensed operators have equal access to available frequencies. This rule changes during emergencies declared by the FCC. A true emergency is defined as a situation that poses an immediate threat to human life or the imminent destruction of property. In these situations, standard operational rules are superseded. Priority access to frequencies may be granted to amateur radio operators to facilitate emergency communications.

Amateur radio communications intended for use in an emergency must meet strict criteria. The communication must directly pertain to the immediate safety of human life or property protection. Furthermore, it must be confirmed that no other reasonable means of communication (such as phone lines or commercial services) was available before or during the event.

Nets and Emergency Operation

In emergency communications and organized radio operations, a net is a group of stations that communicate on a specific frequency. As discussed earlier, these organized communication networks serve various purposes. These nets can be broadly classified into three primary categories.

- **Emergency Communications Nets**: These nets are established to facilitate the relay of critical information during emergencies and disasters. They provide vital communication channels when traditional infrastructure may be compromised.
- **Traffic Nets**: Traffic nets efficiently relay messages between radio operators. They often employ a structured format and procedure to ensure the quick and accurate transmission of information over potentially long distances. Like a postal service, they handle routine communication traffic and coordinate the flow of information across networks.
- **Social or Topical Nets**: These nets provide a platform for amateur radio operators with shared interests to connect and discuss specific topics. They may focus on hobbies, technical aspects of radio operation, or general conversation and camaraderie.

Net Control Stations (NCS)

A net control station (NCS) is the central point of contact within directed nets. The NCS manages the net's operation and ensures orderly communication. In a directed net, all transmissions must be addressed through the NCS. Before you integrate into an active net, listen and observe the operation for some time to familiarize yourself with the protocols and procedures specific to that net.

Amateur Radio Emergency Service (ARES)

The Amateur Radio Emergency Service (ARES) is organized under the ARRL. It is a network of licensed amateur radio operators who volunteer to provide emergency communication support. Membership in ARES is open to any licensed amateur radio operator.

ARES teams function locally and are led by emergency coordinators (ECs). District emergency coordinators (DECs) provide additional leadership in larger geographical areas, such as counties. The ARRL section emergency coordinator (SEC) provides overall coordination and leadership.

The core mission of ARES is to provide communications assistance to local and regional government agencies and relief organizations during emergencies or times of need. This support is vital to disaster response and public safety efforts.

Radio Amateur Civil Emergency Service (RACES)

RACES is a specialized protocol designed to provide emergency communications support to local, county, and state government agencies. As part of the Amateur Service, RACES communications are guided by the regulations established by the FCC. Also, all RACES must function under the express direction of a civil defense organization.

RACES members may routinely train and test without additional special authorization. However, these drills are usually restricted to a maximum of one hour per week. To become active members of RACES, individuals must register with their local civil defense organization. To be a control operator of a RACES station, you must hold a valid FCC-issued amateur radio operator license.

Distress Calls

Receive a Distress Call

When they hear a distress call, hams must cease all transmissions immediately. This ensures that critical information from the station in distress is clear and rescue services have access to the frequency.

- **Acknowledge and Assess**: Immediately acknowledge the station in distress. This confirms the call has been heard and establishes communication. Gather as much information as possible about the nature of the emergency and determine the type of assistance required.

- **Halt Transmissions**: Cease all communications. Open the frequency for emergency transmissions to allow clear communication between the distressed and potential rescuers.
- **Gather Critical Details**: Obtain these vital details from the station in distress: location (the precise location of the emergency), nature of the emergency (for example, medical, fire, or vessel in danger), and the specific assistance required (such as rescue personnel, medical aid, or evacuation).
- **Seek Assistance and Standby**: If you are not in a position to directly assist the station in distress, immediately contact relevant authorities. These could include the coast guard, maritime rescue services, or appropriate emergency response personnel. Transmit the information gathered from the distressed station as accurately as possible.

Send a Distress Call

If you have an emergency during the operation of a radio station, follow the steps below.

- **Signal the Emergency**: On voice modes, transmit "Mayday Mayday Mayday" three times. In Morse code mode, transmit "SOS SOS SOS" three times. Optionally, follow the distress signal with the phrase "Any station, come in, please."
- **Identify Yourself**: Transmit your station's call sign clearly. Repeat this several times if necessary.
- **Provide Location**: State your location with as much specificity as possible. Include latitude, longitude, landmarks, or other details to help pinpoint your position.
- **Nature of Emergency**: Describe the nature of the emergency (for example, medical emergency, fire, or vessel sinking) as concisely as possible.

Assistance Required: Precisely state the type of assistance you need (for example, medical evacuation, search and rescue, or fire suppression). Include any other relevant information that may assist rescuers.

Digital Operation Procedures

Digital Message System Gateways – Winlink

Gateways are specialized stations that bridge radio-based digital messaging networks (like Winlink) with the broader internet. This integration allows people to send and receive emails via amateur radio. Gateways broadcast their operating frequencies. To initiate contact, stations transmit a connection message on the gateway's published frequency. A list of active Winlink gateways and their frequencies can be found on the official Winlink website.

FCC Rules for Digital Operating

The FCC oversees digital mode usage. It specifies maximum allowable bandwidths and data transmission rates. If custom digital codes are used, the FCC mandates that they be made publicly available to ensure transparency within the ham radio community. FCC Part 97 regulations for third-party communications apply equally to digital transmissions and other modes. There are no exemptions for digital communications.

FCC Rules for Automatically Controlled Digital Stations

Stations configured for automatic transmission of RTTY or other data modes are granted expanded privileges to operate on the six-meter band and longer wavelengths. In addition, they may operate within specific segments of the HF bands.

If an automatically controlled station initiates a contact outside of designated automatic control band segments, the station must operate under direct local or remote supervision by a licensed operator. Communication between automatically controlled stations that use RTTY or data emissions is restricted to the non-phone Extra Class portions of the amateur radio bands.

Chapter 3: Radio Wave Propagation

The Ionosphere

HF propagation relies heavily on the ionosphere. This dynamic region of the upper atmosphere, heavily influenced by solar radiation, determines the effectiveness of sky wave propagation. The central focus of this chapter is to explore the nature of the ionosphere and how it impacts radio communications.

Three Ways to Propagate HF Waves

Radio waves can travel over long distances on HF bands via three main methods: sky wave, space wave (line-of-sight), and ground wave propagation.

- **Sky Wave Propagation**: Signals are directed toward the upper atmosphere, where they are refracted (bent) by ionized layers back toward the earth's surface. When an HF signal is beamed upward, it interacts with the ionized layers of the atmosphere. Under the right conditions, these layers act like a giant mirror and refract the signal back toward Earth. This refraction can happen multiple times, known as "hops," and enables transcontinental or even global communication. This allows DX communication and is also known as "skip" or "hop" propagation.

- **Space Wave (Line-of-Sight) Propagation**: In this mode, the transmitting and receiving antennas must maintain a direct, unobstructed visual path. The signal travels in a relatively straight line, which limits its effective range mainly to the horizon. Space wave propagation is generally confined to shorter distances. Line-of-sight is mostly used for VHF and UHF communication but can have some importance at the higher end of the VHF spectrum, such as in local repeater use.

- **Ground Wave Propagation**: Ground wave signals closely follow the earth's curvature (that is, they're propagated along the earth's surface). This propagation mode depends on the surface's electrical properties, especially its conductivity. Ground wave signals weaken as distance increases, and this attenuation is more pronounced at higher frequencies. Within the HF spectrum, ground wave propagation mainly facilitates communication over shorter distances (a few hundred miles or less). It gains significance at lower frequencies (medium frequency and below), where signals can travel substantial distances. Although most prevalent at lower frequencies, ground wave propagation can be involved in certain HF scenarios.

The Ionosphere: A Dynamic Region

The ionosphere is not a clearly defined layer. It is a region within the upper atmosphere that spans a vast altitude range (approximately 30 to 300 miles above Earth). Intense ultraviolet radiation and X-rays from the sun are the main drivers for the creation of ions and free electrons within the ionosphere. This process, where electrons are knocked loose from atoms, is called ionization.

Electron density, and thus how much radio waves are refracted, is not uniform across the ionosphere. Factors that influence this include:

- **Time of Day**: Sunlight is less intense at night, and this causes changes in ionosphere behavior.
- **Solar Activity**: Fluctuations in the sun's output (sunspots, solar flares) can dramatically alter the ionosphere.

Though often simplified as reflection, the ionosphere strongly influences radio wave propagation through refraction. This means the wave's path bends when it encounters regions of different electron densities within the ionized layers. Think of how light bends when it enters water or a glass prism. Radio waves experience a similar effect when they interact with varied electron densities within the ionosphere. Sky wave propagation, the cornerstone of long-distance HF communication, is only possible due to the ionosphere's ability to refract radio waves. The precise amount of refraction determines whether a signal returns to Earth or escapes into space.

Ionosphere Regions and Heights

The ionosphere is not a homogenous region. It exhibits distinct layers characterized by varied electron density and at different altitudes. The intensity of solar radiation and the specific composition of the atmosphere at different altitudes influence the degree of ionization. Consequently, the ionosphere's layers (labeled the D, E, and F regions) form at varied heights across its vast 30- to 300-mile range. Each layer has unique effects on how radio waves, especially on the HF bands, propagate across Earth.

The Transient D Region and Absorption

The D region is the innermost layer of the ionosphere. It is the closest of all the layers to the earth's surface and is situated approximately 30 to 60 miles above the earth's surface. Unlike the higher layers, the D region exists only during daylight hours. Solar ultraviolet radiation ionizes atmospheric molecules to create the D region. After sunset, this ionization rapidly diminishes, and the D region largely disappears.

The D region has a significant absorptive effect on radio signals, especially at lower frequencies. This absorption intensifies with stronger sunlight. On frequencies below 10 MHz, signals can be severely weakened or even completely absorbed during daylight hours. The D region's absorptive properties pose challenges for long-distance communication on lower HF bands (160-, 80-, and 40-meter). Propagation on these bands tends to improve significantly after nightfall when the D region fades away.

The E Region

Located roughly 60 to 70 miles above Earth, the E region shares some similarities with the D region. A product of solar radiation, the E region forms during daylight and weakens at night. However, its electron density is higher than the D region, and this allows it to persist for a somewhat longer period after sunset.

The E region's higher altitude gives it a different impact on radio signals than the lower D region, as it allows for signal refraction rather than absorption. Skips enabled by the E region can occasionally reach up to 1200 miles. Furthermore, the E region is central to several specialized propagation modes, as highlighted below.

- **E-Skip**: Sporadic bursts of ionization within the E region can support short-lived, long-distance communication, mainly on the six-meter VHF band.
- **Meteor Scatter**: The ionized trails left by meteors that enter the E region can briefly reflect radio signals.
- **Polar VHF Enhancement**: Increased ionization of the E region near the earth's magnetic poles can sometimes enable long-distance propagation on VHF bands.

The F Region

The F region is located approximately 100 to 300 miles above the earth's surface. The region is the most complex and influential layer for long-distance HF sky wave propagation. Its behavior is tightly linked to solar activity. It exhibits significant variations throughout the day, across seasons, across latitudes, and with changes in the sunspot cycle.

During daylight hours, the F region splits into two sublayers: the F1 and the higher-altitude F2 layer. This separation is driven by increased solar radiation. At night, these layers usually recombine into a single F region. Several factors, as mentioned above, shape the electron density and height of the F1 and F2 layers.

The F region directly enables some of the longest possible single-hop distances for sky wave propagation, sometimes up to 2,500 miles. Generally, stronger sunlight produces a higher and more ionized F2 region. Stronger sunlight also supports the use of higher frequencies on HF bands.

Radio Wave Refraction

The ionosphere's ability to bend radio waves, known as refraction, is essential for sky wave propagation on HF bands. Imagine a car on a flat surface that suddenly encounters a muddy patch at an angle. The wheels enter the mud first and slow down, which causes the car's path to turn. Similarly, when a radio wave enters a region of the ionosphere with varied electron density, parts of the wavefront encounter different resistance, which causes a change of direction termed refraction.

Two factors influence the extent of this refraction:

- **Electron Density and Ionization Levels**: The ionosphere is full of free electrons created by solar radiation that strips electrons from atoms. The higher the concentration of these free electrons (electron density), the greater the ionosphere's ability to refract radio waves. Think of the electrons as little obstacles. A wave that travels through a region with lots of electrons encounters more resistance on one side of its wavefront. This uneven resistance causes the wave to bend. Since solar radiation is the main driver of ionization, the ionosphere's refractive properties change drastically between day and night. This refractive ability also varies with the 11-year solar cycle.

- **Radio Frequency**: There is an inverse relationship between radio frequency and refraction. The higher the frequency of a radio wave, the less it is refracted by the ionosphere. High-frequency radio waves (like VHF and UHF) interact less with the free electrons within the ionosphere compared to lower-frequency HF waves. Therefore, they tend to pass through with minimal bends. Lower frequencies (within the HF range) are more readily refracted than higher frequencies. This explains why HF bands are so well-suited for long-distance sky wave propagation—they are within the ideal range to experience significant refraction. VHF and UHF, on the other hand, rely predominantly on line-of-sight communication.

Critical Angle and Critical Frequency

Both critical angle and critical frequency are dynamic values. They depend on the varied state of the ionosphere, which is heavily influenced by solar activity and time of day.

Critical Angle

The critical angle refers to the maximum angle of a radio wave that leaves a transmitter. Under a given set of ionospheric conditions, this will still be refracted enough to return to the earth's surface. If the wave exceeds the critical angle, it will not be sufficiently bent by the ionosphere and will escape into space.

To illustrate, imagine a castle under siege. Defenders use a giant catapult to launch heavy stones toward the attackers. The launch angle is vital to reach the enemy lines. There is a maximum angle at which the catapult can launch a stone. If launched too steeply, the stone will not travel far enough and will fall short of the target. A shallow launch angle (low takeoff angle) allows the stone (radio wave) to travel a longer distance within the curved path of the ionosphere before it reaches its target (Earth's surface). This is analogous to successful refraction for sky wave propagation.

Critical Frequency

The critical frequency is the highest frequency that will be refracted back to Earth by the ionosphere when the radio wave enters at a specific angle. Frequencies above the critical frequency tend to pass through the ionosphere rather than refract back down.

To illustrate, picture a high, solid wall over which you need to throw tennis balls (which represent radio waves). You have an assortment of tennis balls with varied amounts of bounciness (which represent different frequencies). Lower-frequency waves would be like less bouncy balls, while higher-frequency waves would be extremely bouncy. There is a maximum height at which you can throw the least bouncy balls (lowest frequencies) and still clear the wall. Higher frequencies, because of their bounciness, can clear the wall even when thrown higher. The maximum height at which you can throw the least bouncy ball represents the critical frequency.

The wall represents the ionosphere and its varied electron density. A higher density is like a thicker, taller wall. The bounciness of the tennis balls represents how much a radio wave interacts with the ionosphere. Less bouncy balls (lower frequencies) are refracted (bounce back) more easily. Radio waves at or below the critical frequency will not experience enough refraction (bounce) to 'clear the wall' and return to Earth. Waves above the critical frequency will pass through the ionosphere like excessively bouncy balls that sail right over the wall. The critical frequency changes based on the "thickness" of the ionosphere, which varies with solar activity and time of day. The "wall" gets taller during periods of intense solar radiation.

Sky wave Propagation, Ground-Wave Propagation, and Skip Zone

Sky Wave Propagation

Sky wave propagation is the mechanism through which radio signals on HF bands can reach distant locations. An HF signal is radiated upward toward the ionosphere. The varied electron density within the ionized layers causes the signal's path to bend (refract), much like light bends when it passes from air into water. With sufficient refraction, the signal is directed back toward the earth's surface. Based on conditions, a signal can refract multiple times within the ionosphere, which can extend its reach even further. Each refraction event is termed a "hop."

The specific ionospheric layer involved significantly influences the maximum possible hop distance. The higher F2 region generally facilitates skips of up to 2,500 miles, while the lower E region supports skips of approximately 1,200 miles. Due to its height and variable behavior, the F2 region provides the longest single-hop distances via sky wave propagation.

Ground Wave

Ground wave propagation refers to radio signals that travel directly along the earth's surface. This mode is influenced by surface conductivity and tends to diminish in strength with distance.

Skip Zone

The skip zone, sometimes referred to as the dead zone, is a region where reliable communication via either ground wave or sky wave propagation is difficult or impossible. The skip zone typifies the limitations of ground wave and sky wave propagation. Reliable communication with a specific station may be difficult or impossible within the skip zone. Stations that attempt to contact each other might find themselves located within each other's skip zones.

Ground wave signals lose strength as they travel along the earth's surface. At increased distances, these signals become too weak to be reliably received. Sky wave propagation depends on signals that refract off the ionosphere and return to Earth at a considerable distance from the transmitting station. There exists a region too far for an effective ground wave but too close for sky wave signals to reflect back down.

Multihop

Multihop propagation refers to a signal that bounces off the ionosphere multiple times to reach its destination. Imagine a flat stone that is skipped across the surface of a water body—each bounce is like a single "hop." Unlike a single hop, multihop allows signals to travel over longer-than-usual distances. Multihop occurs especially on HF frequencies when signals reflect off the ionosphere and back down to Earth and bounce several times before they reach the receiver. This can be useful when there is no clear line of sight between transmitter and receiver due to the curvature of the earth.

Long Path and Short Path

The earth's curvature allows two potential paths for radio signals to travel between distant points: a short path and a long path.

- **Short Path**: In most HF contacts, the most direct route is used—the short path.
- **Long Path**: The alternative, less common route is the long path, where signals travel the long way around and cover a significant portion of Earth's circumference. To attempt to establish a long-path contact, antennas need to be rotated 180 degrees from the short-path direction.

There might be scenarios where long-path propagation succeeds while no short-path signal is detected. Variations in ionospheric conditions along these different paths, along with time of day and seasonal factors, contribute to this possibility.

Effects on Signal Reception

Due to the vast distances involved, long-path propagation can produce some unique effects on received signals. Two notable phenomena that result from it are delayed echo and self-reception.

- **Delayed Echo**: This is essentially a case of multiple paths and multiple arrivals. The short-path and long-path routes represent vastly different distances. If ionospheric conditions allow signals to propagate successfully along both paths, the signal that travels the long path will arrive at the receiver slightly delayed compared to the short path signal. This time difference results in the perception of a faint echo shortly after the primary signal. The degree of delay and the strength of the echo depends on the relative strength of the signals that arrive via each path.
- **Self-Reception**: Transmitters generally do not hear themselves on the same frequency. However, under exceptionally strong propagation conditions, a

powerful signal can circle the entire globe via the long path and be faintly detected by the transmitting station. Due to the speed of radio waves and the earth's circumference, this self-received signal will be delayed by approximately one-seventh of a second from the original transmission.

Consider communication between the eastern United States (for example, North Carolina) and Japan. The short path (which travels westward over the Pacific Ocean) covers approximately 7,500 miles. Alternatively, the long path takes signals eastward across the Atlantic, Europe, and Asia (roughly 17,400 miles to Japan). These phenomena are not always present in long-path propagation but depend heavily on ionospheric conditions and signal strength.

The Sun

The sun exerts a profound influence on the ionosphere and, consequently, HF radio propagation. Its activity exhibits a cyclical pattern known as the solar cycle, with an average length of roughly 11 years.

Sunspots

Sunspots, dark regions with powerful magnetic fields, are a useful indicator of overall solar activity. A greater amount of sunspots corresponds to increased levels of ultraviolet (UV) radiation emitted by the sun. Increased solar activity, as indicated by higher sunspot numbers, causes stronger ionization of the upper atmosphere. This enhanced ionization generally favors HF propagation, especially on the higher frequency bands where increased refraction supports longer communication distances.

Solar Activity

The approximately 11-year solar cycle causes significant changes in ionospheric conditions. During periods of low solar activity, the higher HF bands (15-, 12-, and 10-meter) become less reliable for long-distance communication due to decreased ionization. The 20-meter band often provides worldwide propagation during daylight hours, even at low points in the solar cycle. This makes it a valuable band for reliable communication.

The sun's rotation, with a period of approximately 25 to 27 days, causes periodic fluctuations in the radiation that reaches Earth. This causes regular variations in ionospheric conditions, which directly affect radio propagation across the HF bands.

Atmospheric noise, commonly referred to as "static," tends to be more prevalent during summer months. This increased noise is particularly noticeable on lower HF bands, which could potentially mask weaker signals and hinder communication.

Measure Solar Activity

Certain indices are regularly monitored to gauge how the sun's activity affects radio propagation. The important ones are highlighted below.

- **Solar Flux Index (SFI)**: Expressed in solar flux units, the SFI measures the intensity of solar radio emissions at a wavelength of 10.7 cm (2800 MHz). This index correlates well with sunspot activity.
- **K-Index (of Geomagnetic Stability)**: The K-index provides a measure of fluctuations in the earth's magnetic field. It ranges from 0 (quiet) to 9 (extreme storm), with each value representative of a three-hour average.
- **A-Index**: The A-index provides a long-term perspective on geomagnetic field stability on a global scale. It is a daily average calculated from eight individual K-index readings. Values range from 0 (stable) to 400 (highly disturbed).

Assess Propagation

Two critical frequencies help radio operators understand the range of usable frequencies for a given communication path. These frequencies are maximum usable frequency (MUF) and lowest usable frequency (LUF).

Maximum Usable Frequency (MUF)

MUF represents the highest frequency that will be refracted back to Earth by the ionosphere at a given time and for a specific path. Generally, frequencies that exceed the MUF will not be adequately refracted by the ionosphere and tend to penetrate and escape into space. To illustrate the effects of the MUF on signal transmission, let us assume that the MUF is 20 MHz. The table below further exemplifies this.

Frequency (MHz)	Behavior with Respect to MUF	Notes
10.	Below the MUF.	The signal is effectively refracted back toward Earth and provides reliable sky wave propagation.
20.	Near the MUF.	The signal is refracted but with minimal margin. Propagation might be unreliable.
22.	Above the MUF.	The signal penetrates the ionosphere and escapes into space.

Lowest Usable Frequency (LUF)

LUF marks the lower limit of usable frequencies for a specific path. Frequencies below the LUF will experience excessive attenuation within the ionosphere. This attenuation weakens or completely absorbs the signal before it can reach the intended destination. To illustrate the effects of the LUF on signal transmission, let us assume that the LUF is 10 MHz. The table below further exemplifies this.

Frequency (MHz)	Effects on Signal	Notes
8.	Absorbed by ionosphere.	At 8 MHz, the signal likely undergoes significant absorption within the lower ionospheric layers. This prevents a successful return to Earth.
10.	Barely refracted.	Signals on this frequency experience weak refraction, which could cause unreliable communication or insufficient signal strength.
12.	Refracted to Earth.	Frequencies around 12 MHz and above generally experience sufficient refraction by the ionosphere. This creates the potential for successful sky wave propagation.

Optimum Working Frequency (OWF)

Frequencies between the LUF and MUF provide the best chance for successful sky wave propagation over a particular path. Signals within this range experience the optimal amount of refraction for their frequency.

This OWF should fall slightly below the current MUF but also simultaneously remain above the LUF. To ensure that the signal is refracted back to Earth with sufficient strength, operate slightly below the MUF. To minimize signal absorption within the ionosphere and allow the signal to reach the intended destination, remain above the LUF.

LUF Above MUF: Propagation Blackout

Under certain conditions, the LUF can temporarily rise above the MUF. This situation causes a communication blackout, as no frequencies within the normal HF range will be refracted back to Earth. This phenomenon temporarily halts the usual sky wave propagation on HF bands.

Solar flares are common causes of this phenomenon. Increased ionization in the lower atmosphere produces a highly absorptive ionosphere. This raises the LUF and causes even higher HF frequencies to be attenuated rather than refracted.

Factors that Influence the MUF

The MUF for a given communication path is dynamic and influenced by several factors:

- **Path and Location**: The longer the signal path, the more hops (refractions within the ionosphere) might be required for successful communication. Each refraction introduces some energy loss. Shorter paths generally support the use of higher frequencies (higher MUF) before this attenuation becomes excessive. The MUF also varies based on the regions the signal path traverses. Factors like the average electron density of the ionosphere above the path and the angle at which the signal enters the ionosphere differ across different latitudes and longitudes.

- **Time and Season**: The sun's position in the sky changes throughout the day and seasonally alters the intensity and angle of solar radiation that reaches specific parts of the earth. Consequently, the electron density in the ionosphere, which dictates refraction, exhibits significant diurnal (daily) variations. Usually, the MUF is highest around midday when solar radiation is most intense.

- **Solar Activity**: Increased solar activity, such as sunspots and solar flares, releases intense bursts of ultraviolet and X-ray radiation. This dramatically alters the ionosphere. Sometimes it can raise the MUF or cause disturbances that can destabilize HF propagation.

To minimize signal loss during long-distance sky wave propagation, select an operating frequency slightly below the current MUF. For instance, if the MUF is 15 MHz, the 17-meter band (18 MHz) might produce good results.

Assess Propagation Conditions

Radio operators can use several resources to evaluate current propagation on desired HF bands, as discussed below.

- **Automated Receiver Stations**: Online receiver networks provide a valuable tool for radio operators to assess HF propagation conditions. These networks consist of geographically distributed receiving stations connected to the internet. When they monitor where a specific transmission is detected, operators can gain real-time insights into which band paths are currently open and the extent of their signal's reach.

- **NCDXF Beacon Network**: The Northern California DX Foundation (NCDXF) operates a network of beacons that transmit on bands between 20 and 10 meters. Tuning in to these beacons provides insights into current band openings and propagation paths.

Predicting Sky Wave Propagation

- **Beacon Observation**: Short-skip activity on the 10-meter band often suggests potential openings on higher-frequency bands like six meters. This is because similar ionospheric layers influence propagation on these bands.

- **Propagation Software**: These software programs use complex algorithms to generate propagation predictions. They take into account the time of day and season, current solar conditions (sunspot numbers, solar flux), start and destination locations, and desired frequency band.

- **On-Air Exploration**: When you transmit a general call (CQ) on a band and listen for responses, this can provide a practical, real-time assessment of current propagation. The FT8 digital mode, due to its sensitivity to weak signals, can provide a way to visualize the reception of transmissions across the globe. Stations across the globe might hear and respond to an FT8 transmission, even if voice or CW communication is not feasible.

FT8 as a Propagation Indicator

To estimate which bands are open, you can listen on the appropriate FT8 frequency, as presented in the following table.

HF Band	FT8 Frequency (MHz)
160 meters.	1.840.
80 meters.	3.573.
40 meters.	7.074.
30 meters.	10.136.
20 meters.	14.074.
17 meters.	18.100 (The 17-meter FT8 frequency might not be an official allocation. Always check your local band plans and regulations.)
15 meters.	21.074.
12 meters.	24.915.
10 meters.	28.074.
6 meters.	50.313.

Ultraviolet, X-ray, and Solar Particles

The sun's activity can significantly disrupt radio communication, especially on HF bands. The main types of solar disturbances and their effects are discussed below.

Light-Speed Disturbances

Solar flares arrive at Earth within approximately eight minutes. They dramatically increase D-region ionization and cause severe signal absorption and potential radio blackouts that last from seconds to hours. Lower HF frequencies are generally affected more than higher ones, and the impact is mostly limited to the sunlit side of the earth.

Particle Disturbances

The sun constantly emits charged particles that form the **solar wind**. This alone can influence radio propagation conditions. However, more dramatic disruptions in HF communications stem from two main sources:

- **Coronal Holes**: These "weak spots" in the sun's magnetic field allow a continuous outflow of charged particles to escape. Coronal holes contribute to the solar wind and enhance its intensity.
- **Coronal Mass Ejections (CMEs)**: When large eruptions occur on the sun, immense quantities of charged particles and solar material are propelled outwards. CMEs are slower than light and can take anywhere from 15 hours to several days to reach Earth.

Geomagnetic Storms

Charged particles from solar disturbances, such as coronal holes and CMEs, interact with Earth's magnetic field. They bombard the E region of the ionosphere, which causes a significant increase in ionization within this layer.

This interaction also triggers geomagnetic storms—temporary disturbances characterized by fluctuations within the earth's magnetic environment. Geomagnetic storms destabilize the ionosphere and have an adverse impact on HF communication across all latitudes, especially in high-latitude regions. Propagation can degrade severely or become impossible for extended periods that can range from hours to days. Conversely, geomagnetic storms can also create auroras, which are beneficial to VHF transmissions.

Auroras: Benefits of Geomagnetic Activity

As charged particles are guided by Earth's magnetic field toward the poles, they collide with atmospheric gases, which causes them to become ionized and excited. When these ionized gases return to their normal energy state, they release energy as light and create the colorful display known as an aurora.

Although geomagnetic activity disrupts HF bands, auroras can sometimes reflect VHF signals. This reflection is common on the six-meter and two-meter bands. Auroral propagation is often accompanied by a distinctive "fluttery" quality on received signals.

Scatter Modes

Sky wave communication relies on the ionosphere's ability to refract radio waves and direct them back toward Earth for long-distance communication. However, there is another mechanism that can sometimes bridge communication gaps—scatter propagation. Signal scatter can extend a station's reach into areas normally located within the skip zone.

<u>What is Scatter?</u>

Imagine a large, mirrored disco ball suspended above a dance floor. This represents a radio transmitter that sends out a signal. The disco ball represents irregularities within the atmosphere. When the disco ball is illuminated with a spotlight (the radio signal), some light beams directly reach the dance floor below. This represents ground wave propagation.

As the disco ball spins, light beams are reflected in countless directions, and this scatters light throughout the room. This scattered light represents scatter propagation, where radio waves are dispersed by irregularities. Some corners or areas too far from the direct spotlight might still be faintly lit due to the scattered light that bounces off the disco ball. Equally, areas within the skip zone can sometimes receive weak, scattered signals. The lights can also reflect off the disco ball back to the spotlight, similar to how backscatter works.

The light from the disco ball that hits the walls and corners is significantly weaker than the directly illuminated floor, just as scatter signals tend to be weak. In addition, the light might appear stretched or distorted as it bounces off the disco ball's many facets. This closely mirrors how various paths can produce distortion in scatter propagation.

Furthermore, just as the reflections on a disco ball constantly shift as it spins, scatter propagation is somewhat unpredictable due to the variable nature of atmospheric irregularities. The disco ball adds a dynamic and visual element and demonstrates how scatter can extend a signal's reach into unexpected areas.

Characteristics of Scatter

Scatter propagation provides a way for signals to extend their reach. It enables some communication within the skip zone—the area where ground waves are too weak for reception and distances are too short for traditional sky wave refraction.

Backscatter

Backscatter is a specific type of scatter where a portion of the scattered signal is directed back toward the transmitting station's general direction. This phenomenon is important for communication within the skip zone.

The skip zone is an area too far for consistent ground wave reception and too close for signals to be fully refracted back down by the ionosphere. Backscatter can enable reception from stations within this skip zone. It redirects some signal energy back toward the transmitter, which extends the signal's reach into areas normally unreachable by main propagation modes.

Weak Signals

Only a tiny fraction of a transmitted signal's energy actually gets scattered in the right direction to reach a receiver within the skip zone. Consequently, scatter signals are often extremely weak and require sensitive receivers and careful listeners to detect.

Flutter Signals

The main culprit behind the flutter is the dynamic nature of the regions where scatter occurs. The ionosphere is not a perfectly uniform layer. Electron density, which influences how much radio signals bend, constantly varies due to solar activity and other factors. These changes in the scattering medium affect how the signal propagates.

The scattered signal travels like ripples on a pond. However, this "pond" (the ionosphere and troposphere) is constantly disturbed by new waves, wind patterns, and other changes. These disturbances cause the scattered signal path to continuously strengthen or weaken.

These constant path fluctuations cause the strength of the signal that arrives at the receiver to fade in and out rapidly. This translates to wavers or flutters when the signal is demodulated into audio.

Distortions Due to Multipath

Scattered signal energy often reaches the receiver's antenna via multiple slightly different paths due to the numerous irregularities it encounters. When these various signal components arrive slightly out-of-sync, they can interfere with each other. This multipath interference is a major cause of the distortion commonly observed in scatter signals.

Near Vertical Incidence Sky Wave (NVIS)

Near Vertical Incidence Sky Wave (NVIS) is a propagation mode in the MF and HF bands in which signals are transmitted at very steep (high elevation) angles toward the ionosphere. This technique enables reliable short-distance communication, usually within a few hundred kilometers of the transmitting station.

When signals are directed nearly vertically, they encounter the ionosphere and undergo a different type of refraction and scatter compared to lower-angle sky wave propagation. These signals are scattered back toward Earth and cover a relatively localized area around the transmitter.

NVIS is most effective on the lower HF bands, especially the 40-meter band and below. This is because these frequencies are more likely to be reliably reflected by the ionospheric layers (primarily the F region), which are essential for NVIS to function.

A horizontally oriented dipole antenna positioned close to the ground (approximately one-eighth to one-quarter wavelength above ground to minimize ground losses) is optimal for NVIS propagation. A horizontal dipole radiates equally in all directions around it (horizontally), which broadcasts the NVIS signals in a circular area around the transmitting station. Furthermore, this type of antenna radiates most of its energy at high elevation angles and can provide coverage over a localized region.

Chapter 4: Amateur Radio Practices

HF Station Installation

The configuration of your ham radio station is completely your personal choice, from the equipment you select to how you arrange and connect components. There is room to experiment and tailor the setup to your specific style and preferences. However, there are fundamental principles that must be followed. This section examines these principles and how they apply to amateur radio.

RF Interference (RFI)

When you operate a radio transmitter, this inevitably creates the potential for radio frequency interference (RFI). This occurs when your transmitted signals inadvertently disrupt the normal operation of nearby electronic devices, such as radios, televisions, computers, and various appliances.

Intermodulation

Intermodulation (IMD) occurs when two or more strong RF signals encounter a component or connection that does not have a perfectly linear response (like a corroded connector or even a rusty bolt). This nonlinear element acts like an unwanted mixer and creates new signals at sums and differences of the original frequencies and their harmonics. These intermodulation products (IMPs) can fall on the frequencies you want to use and cause interference. For example, if you transmit on 14.200 MHz and there is a strong signal on 14.100 MHz, IMD in a rusty fence nearby could create interference.

To solve the problem of intermodulation, follow the steps below:

- **First Suspect**: If interference suddenly appears when you transmit, IMD is a likely culprit.
- **Track It Down**: IMD sources are often external to your station. Directional antennas and careful investigation can help you pinpoint the source.
- **Fix the Root Cause**: Clean up the rusty connection or nonlinear element. A transmission filter will not help since the interference is generated externally.

Arcing

Arcing occurs when an electrical current jumps across a gap in a conductor, usually due to high voltage or a poor connection. This creates a bright spark or sustained arc, which can be a significant source of RFI. The rapid on-and-off nature of the arcing creates a wide range of electromagnetic noise across the radio spectrum. This 'broadband' noise can disrupt reception across a wide range of frequencies.

The solution requires you to first identify the location of the arc. Visually inspect power lines, connections in your station, and any nearby equipment that might be the reason for the problem. A filter will not eliminate the interference because arcing generates broad-band noise. The definitive approach is to address the source of the arcing. You might need to tighten loose connections, have faulty power lines repaired by a qualified electrician, or ensure proper maintenance of motors.

Harmonics

RF amplifiers, even well-designed ones, can introduce non-linearities. These non-linearities cause a portion of the input signal's energy to generate new signals at multiples of the fundamental frequency. For example, if you transmit on 7.150 MHz, your transmitter might also generate weaker signals at 14.300 MHz (second harmonic) and 21.450 MHz (third harmonic). Close-by radios and TVs may be tuned to frequencies near these harmonics and experience unwanted noise or even complete signal blockage.

The solution to harmonics RFI is a low-pass filter. This filter is inserted with your transmission line (feed line) before it reaches your antenna. The filter allows your desired operating frequency and a small margin above to pass through, while it significantly attenuates signals at higher frequencies, such as harmonics. For optimal performance, the low-pass filter's impedance should be matched to the characteristic impedance of your feed line. This ensures minimal signal loss at your desired operating frequency.

Other Types of RF Interference

Direct pickup means your RF signal is directly received by a device's internal wires. **Common-mode** interference occurs when RF travels on the outside of cables and disrupts the electronics they connect. **Fundamental overload** happens when a strong signal simply overwhelms a nearby receiver's ability to filter the desired signal.

Vehicle-Based RFI Sources

Modern vehicles are packed with electronics that can generate radio frequency interference. Common culprits include:

- **Fuel Pumps**: Electric fuel pumps, often located within the fuel tank, contain electric motors that can generate broadband noise across a wide range of frequencies. This noise can couple into your vehicle's wires and can interfere with your radio.

- **Onboard Computers**: Vehicles rely on numerous computer systems for engine control, navigation, and entertainment. The high-speed digital signals within these computers can create RF emissions that might disrupt radio reception.
- **Engine Electrical Systems**: The ignition system (spark plugs and related wires), alternator, and various electric motors in the vehicle can also be sources of RFI. Arcing contacts and worn components often exacerbate the problem.

How to Identify RF Interference

When a strong RF signal from transmissions interferes with audio equipment, the radio operator will likely encounter symptoms like crackles, pops, or snaps timed with transmissions. With AM phone signals, the affected device might reproduce a distorted version of the operator's voice. On SSB, the voice will sound unintelligible.

Eliminate RF Interference

To minimize the chance of RFI within the ham shack, consider the following strategies:

- **Grounding and Bonding**: Think of grounding and bonding as a unified electrical team for your station's equipment. When you interconnect the metal enclosures of your transceiver, amplifier, and other devices with heavy-gauge wire, you establish a "common ground" reference point. This minimizes any voltage differences between equipment chassis that could create unwanted currents and cause RFI. A proper connection to an external ground rod (driven deep into the earth) further reduces the station's susceptibility to strong external signals that might otherwise couple into your equipment. In automobiles, the vehicle chassis often provides sufficient grounding. However, additional grounding can be improved by connecting the radio's ground to the battery's negative terminal or the ground strap.
- **Shielded Cables**: Imagine that your DC power cables and signal cables act like tiny antennas. Unshielded cables can pick up stray RF signals from the environment, and conversely, they can radiate some of your station's RF energy. Shielded cables address both concerns. A braided metal shield surrounds the inner conductor and blocks external RF from the signal wires inside. This also helps contain your station's transmitted signals within the cable and prevent them from interference with nearby electronics.
- **Ferrites**: These doughnut-shaped ferrite cores can be placed on cables to act as RF "choke points." The ferrite material has high impedance at radio frequencies and hinders the flow of common-mode RF currents. Imagine a common-mode current as water that struggles to get through a sponge. The ferrite core makes it

difficult for the RF energy to travel along the cable sheath (common mode) and potentially enter or exit the station.

- **Antenna Considerations**: The closer your antennas are to your ham shack, the higher the chance your equipment will directly pick up strong RF signals. Maintain a reasonable distance between your antennas and the operating position to mitigate this. Balanced antennas have another advantage. Unlike unbalanced antennas (where the feed line carries both the signal and the ground return current), balanced antennas have a symmetrical design that minimizes the chance of RF currents in the feed line or power connections, which further reduces the potential for RFI within the station.

- **Bypass Capacitors**: These components are useful to suppress RF within audio circuits and control lines. As they provide a low-impedance path for RF to ground, they help keep RF from sensitive sections of devices where it can cause distortion or erratic operation. Do not place bypass capacitors across speaker outputs or data lines, as this can negatively affect the intended signal.

Mobile Power and Antenna Installation

For optimal performance of a 100 W HF radio in your vehicle, connect it directly to the battery with a thick, heavy-gauge wire. This minimizes voltage drops that could cause the radio to malfunction. Always include fuses in both battery leads as a safety measure. Do not use a vehicle's auxiliary power outlet. They are designed for low-current devices and cannot handle the demands of a high-power transceiver. Vehicle antennas are often smaller and less efficient than the ones used for home stations.

Basic Test Equipment

Radio and antenna systems can present occasional technical challenges. A ham must be able to troubleshoot potential issues, optimize the antenna system, and experiment with different radio configurations. A few tools are used to diagnose problems, verify performance, and enhance overall station setup.

Analog and Digital Multimeters

A multimeter is indispensable for any electronics enthusiast, especially amateur radio operators. This device is used to measure voltage, current, and resistance. It can also be used to test diodes, transistors, and the continuity of electrical connections between two points. There are two types of multimeters:

- **Analog multimeters** display measurements with a needle that moves across a graduated scale. This provides a better visual representation of gradual changes and is preferred when circuits are fine-tuned where a maximum or minimum value is desired.

- **Digital multimeters** display measurements numerically on a digital screen. They generally have superior precision for most measurements. They often include features that enhance their versatility:
  - **Auto-ranging** selects the optimal measurement range automatically.
  - **Peak-hold** captures and displays the maximum value encountered. It is useful for transient measurements.
  - **Extended functions**: Some digital meters can measure frequency, capacitance, temperature, and other parameters.

When you take voltage measurements, select a meter with a high input impedance. This minimizes the meter's impact on the circuit under test, a phenomenon known as "loading." A meter with excessively low input impedance can draw significant current, which can alter the voltage measured and cause inaccurate results.

Oscilloscope

Oscilloscopes, often called "scopes," provide a graphical representation of voltage variations over time. They analyze and display complex audio, data, and radio waveforms. Oscilloscopes use horizontal and vertical channel amplifiers, along with variable gain and update rate adjustments, to accommodate a wide range of signal types and amplitudes.

Monitor CW Keying Waveform with Oscilloscopes

Oscilloscopes are the most suitable instruments to analyze and troubleshoot the keying waveforms of CW transmitters. Raw transmitter output can overload and damage an oscilloscope and must be attenuated before it is connected. To visualize the modulation envelope of transmitted signals, connect the attenuated transmitter output to the vertical channel of the oscilloscope.

An oscilloscope highlights subtle aspects of the transmitted signal that basic numeric meters cannot measure. This visual feedback can be used to modify Morse code keying waveforms, microphone gain levels, and speech processing settings.

Measure Impedance and Resonance with Antenna Analyzers

Antenna analyzers are specialized instruments that combine several functions to facilitate comprehensive antenna and feed line analysis. They display graphical impedance waveforms that can be used to determine resonance and the point of lowest SWR. They usually incorporate several components:

- **RF Signal Generator**: This produces a low-level test signal for transmission measurements.
- **Frequency Counter**: This ensures precise frequency measurement.
- **SWR Bridge**: This determines the standing wave ratio (SWR) on the feed line. First, disconnect the antenna from the radio and connect it to the analyzer. The analyzer then uses small signals from the signal generator to measure the SWR.
- **Impedance Meter**: This measures the complex impedance of the antenna system. In addition to impedance measurements, some advanced analyzers can determine coaxial cable parameters such as velocity factor and electrical length.

Strong signals from nearby transmitters can interfere with antenna analyzer measurements. For optimal accuracy, measurements should be taken in an environment with minimal external RF interference.

VSWR Measurement with Directional Wattmeter

Voltage Standing Wave Ratio (VSWR) is a numerical representation of how well an antenna is matched to its feed line and the radio transmitter. Directional wattmeters measure both the forward power (which travels from the transmitter toward the antenna) and the reflected power (caused by impedance mismatches). The VSWR can be calculated from these measurements.

Field Strength Meters

A field strength meter detects and measures the relative strength of the RF field emitted by an antenna. Knowledge of the field strength is useful to optimize antenna positions and make transmitter adjustments. Field strength can also help map the radiation pattern of an antenna with measurements at various points around the antenna.

Audio Processing

Speech processing techniques aim to enhance the intelligibility of voice transmissions, especially in difficult signal conditions like fade, noise, or interference. This is valuable in AM and SSB modes where, compared to CW, the average power of the voice signal is lower.

Audio speech processing selectively increases the average power of the voice signal without the introduction of excessive distortion. This improves the received signal's clarity under compromised conditions. Common tools include compression (which evens out the loudness of speech) and audio frequency equalization (which selectively emphasizes certain frequencies in the voice signal).

Excessive processing can introduce distortion that may spread a signal into adjacent channels and interference. Processing can sometimes amplify background noise, which can further reduce intelligibility.

The S-Meter

The S-meter on a receiver provides a relative indication of incoming signal strength. Each S-unit represents a significant jump in signal strength:

- **6 dB Change**: When you move up one S-unit, this signifies a fourfold increase in received signal power. Conversely, a signal that drops one S-unit is four times weaker.

The S-meter on your receiver provides a relative indication of incoming signal strength in a linear manner. A moderately strong signal generally registers as S-9 on the meter. However, above S-9, the markings are no longer linear. When the S-meter jumps from S-9 to "S-9 + 20 dB," the received signal is not simply 20 times stronger. It is actually 100 times more powerful. The "dB" markings refer to decibels, a logarithmic scale used to express power ratios. Each 6-dB increase represents quadrupled signal power. So, an "S-9 + 40 dB" signal is 10,000 times stronger than a signal that registers as an S-9.

Correct Sideband Placement

In SSB operation, the radio's displayed frequency represents the carrier frequency, not the actual transmitted voice signal. For USB, the transmitted voice signal occupies frequencies entirely above the displayed carrier frequency. For LSB, the transmitted voice signal occupies frequencies entirely below the displayed carrier frequency.

Responsible operators must ensure their entire SSB sideband remains within their allocated frequency privileges. Since SSB signals commonly have a bandwidth of around 3 kHz, it becomes necessary to allow this 3 kHz buffer zone near the band edges. This bandwidth allowance prevents interference with other bands and frequencies.

For example, assume that the upper band limit for a General Class license holder is 7.300 MHz. When USB is used, the displayed carrier frequency should be set at least 3 kHz lower (to 7.297 MHz). Conversely, when LSB is used, the displayed carrier frequency should be set at least 3 kHz lower (to 7.297 MHz). This placement ensures that the sideband remains within the allocated band.

Alternative Power

Alternative power sources and energy storage solutions are vital for amateur radio operators, especially those involved in portable operations, emergency communications, or off-grid activities. In conjunction with reliable energy storage methods like batteries, solar and wind power systems can provide independence from traditional power grids. Two prominent options in this field include:

- **Solar Power**: Solar energy systems use photovoltaic cells to achieve direct conversion of sunlight into usable electricity.

- **Wind Power**: The large, propeller-like blades of wind generators convert the kinetic energy of wind into rotational mechanical energy. This energy, in turn, drives generators to produce electricity.

Energy Storage for Alternative Power Sources

Alternative power sources like wind and solar are inherently intermittent. This intermittency necessitates energy storage solutions to maintain reliable power. These systems store excess energy produced during peak periods (for example, windy days and hours of strong sunlight) for use later when generation is low or absent.

Batteries are the most widely used energy storage technology for small- and large-scale renewable energy setups. Their ability to store and release electrical energy on demand is essential.

These alternative power sources have some limitations. Periods without wind directly limit power generation from wind turbines, while reduced or absent sunlight (such as at night or during heavy overcast) hinders power production from solar panels. These constraints highlight the importance of energy storage and strategic power plans within alternative energy systems.

Solar Cells

Solar cells are specialized semiconductor devices that directly convert light energy (photons) into electrical energy. Their construction and operation closely resemble that of diodes. When sunlight strikes the solar cell, absorbed photons excite electrons and establish a voltage across the cell's internal junction. This voltage is called the open circuit voltage (the maximum potential difference a solar cell can produce when no current flows). It is usually around 0.5 V DC for a single cell and can be used to power electrical devices.

Configuration of Solar Cells

Multiple individual solar cells are often connected in series within a solar panel to achieve practical voltage levels. For example, a common 12 V solar panel might contain 36 cells connected in series. Furthermore, solar panels frequently employ series-parallel configurations to increase both voltage and current capacity. The series connections increase voltage while the parallel connections maximize current.

Because solar panels can be vulnerable to undesirable discharge or damage, the batteries connected to them must be safeguarded. Two common strategies used to achieve this are discussed below.

- **Block Diodes**: Incorporate a series diode between the solar panel and the storage battery to prevent reverse current flow from the battery back into the panel during low-light conditions. This protects the battery from discharge at night or when there is very little sunlight.

- **Charge Controller (for Lithium Iron Phosphate Batteries)**: Advanced battery chemistries, such as lithium iron phosphate, require a charge controller. This specialized device regulates the process from the solar panel and ensures the battery receives the correct voltage and current. Charge controllers prevent potential damage to the battery.

Chapter 5: Electrical Principles

This chapter discusses vital concepts, such as reactance, resonance, impedance, and the function of active components like transistors and diodes.

Power, Decibels, and Electrical Review

Power

We can draw a useful analogy between the concept of water flow and the fundamental concepts of electrical power, decibels, and circuit behavior. Imagine a fireman who directs water toward a fire. Just as an increase in the size of a hose or the water pressure will deliver more water to the fire, specific electrical adjustments can increase the power delivered within a circuit.

The ability to effectively extinguish the fire depends upon the volume of water that flows through the hose (current) and the force with which the water is expelled (electromotive force or voltage). A larger hose reduces resistance to water flow, analogous to a conductor with lower resistance allowing more electrical current to flow in a circuit. Resistance is measured in ohms and is represented by the symbol R or Ω. Lower resistance permits a greater flow of electrical current, measured in amperes (amps) and symbolized by I. Increased water pressure is analogous to increased voltage (or electromotive force) within a circuit. Voltage, or electromotive force, drives electrical current. It is measured in volts and is represented by the symbol V.

Electrical Units

Knowledge of the fundamental units of measurement in electricity is central to circuit analysis, so calculations can be performed.

- **Current (I)**: This represents the flow of electrons through a conductor and is measured in amperes (A). This measurement is obtained with an ammeter.
- **Voltage (E)**: Also known as electromotive force (EMF), voltage refers to the electrical potential difference that drives electrons through a circuit. It is measured in volts (V) with a voltmeter.
- **Resistance (R)**: Resistance indicates a material's opposition to current flow. It is measured in ohms (Ω) with an ohmmeter.
- **Power (P)**: Power is the rate at which electrical energy is transferred or consumed within a circuit. It is measured in watts (W) and calculated with the

formula P = I × E (Power = Current × Voltage). A wattmeter is used for direct power measurements.

The Magic Circles

Ohm's Law (EIR)

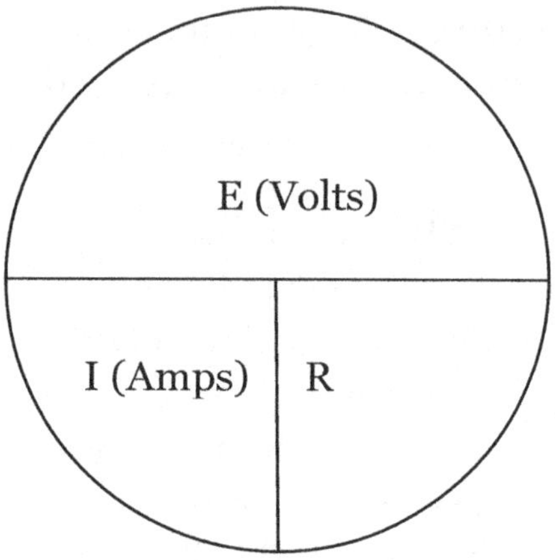

Power Formula (PIE)

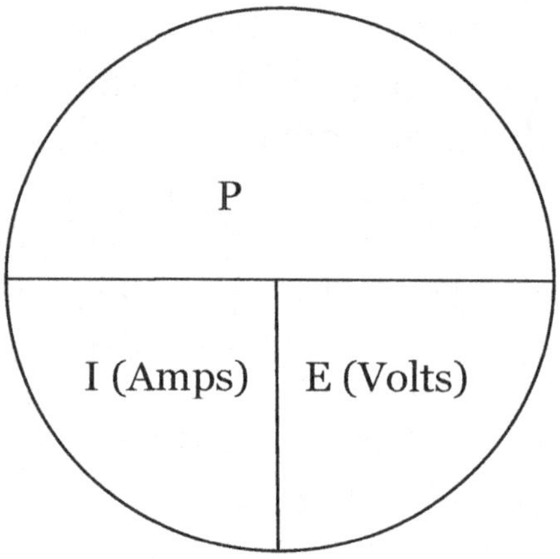

Power Formulas that Involve Resistance

Although power can be calculated directly from voltage and current (P = I × E), it is also possible to derive power calculations when resistance values are known. These formulas are useful when resistance is a primary factor, or direct current and voltage measurements are difficult to obtain. The possible scenarios are highlighted below:

- To calculate power when current and resistance are known: P (Watts) = $I^2 \times R$.
- To calculate power when voltage and resistance are known: P (Watts) = $E^2 \div R$.
- To calculate voltage when power and resistance are known: E (Volts) = $\sqrt{(P \times R)}$.

Power Formula Examples

Example 1: Determine the electrical power consumed by a 160-ohm resistor when supplied with a 200 V DC voltage.

- Required value: Power consumption in watts (W).

- Given information: Voltage (E) = 200 V and Resistance (R) = 160 ohms (Ω).
- Applicable formula: $P = E^2 \div R$.
- Substitute applicable values: $P = 200^2 \div 160 = 250$ W.

Example 2: Calculate the electrical power consumed by a 24 V DC motor with a current draw of 0.5 amperes.

- Required value: Power consumption in watts (W).
- Given information: Voltage (E) = 24 V and Current (I) = 0.5 A.
- Applicable formula: $P = I \times E$.
- Substitute applicable values: $P = 0.5 \times 24 = 12$ W.

Example 3: How many watts are dissipated when a current of 12.0 mA flows through a 2.50 kilo-Ohms resistor?

- Required value: Power consumption in watts (W).
- Given information: Resistance (R) = 2.50 kΩ (= 2500 Ω) and Current (I) = 12.0 mA (= 0.012 A).
- Applicable formula: $P = I^2 \times R$.
- Substitute applicable values: $P = 0.012^2 \times 2500 = 0.36$ W = 360 mW.

Radio Units and Conversion

Prefix	Symbol	Power of 10	Value Examples
Giga	G.	10^9.	Gigahertz (GHz), Gigawatts (GW).
Mega	M.	10^6.	Megahertz (MHz), Megaohms (MΩ), Megawatts (MW).
Kilo	k.	10^3.	Kilohertz (kHz), Kiloohms (kΩ), Kilowatts (kW).
(unit)	-	10^0.	Hertz (Hz), Ohms (Ω), Watts (W), Volts (V).
Milli	m.	10^{-3}.	Millivolts (mV), Milliamps (mA), Milliwatts (mW).
Micro	μ.	10^{-6}.	Microvolts (μV), Microamps (μA), Microfarads (μF).
Nano	N.	10^{-9}.	Nanofarads (nF).
Pico	P.	10^{-12}.	Picofarads (pF).

Examples of Conversion

Example 1: Convert 50 kHz to Hertz (Hz).

- "Kilo" means 1,000.
- So, there are 1,000 Hz in every kHz.
- Multiply the kilohertz value by 1,000: 50 kHz × 1,000 = 50,000 Hz.

Example 2: Convert 150 microvolts (μV) to millivolts (mV).

- "Micro" is smaller than "milli."
- There are 1,000 μV in every mV.
- Divide the μV value by 1,000: 150 μV ÷ 1,000 = 0.150 mV.

Example 3: Convert 2,500 Megawatts (MW) to Gigawatts (GW).

- "Giga" is bigger than "Mega."
- There are 1,000 MW in every GW.
- Divide the MW value by 1000: 2500 MW ÷ 1000 = 2.5 GW.

Decibels (dB)

The decibel (dB) is a logarithmic unit that expresses a ratio between two power levels. It provides a convenient way to represent both increases (gain) and decreases (loss) in signal power within electronic and radio systems. For your exam, focus on these key relationships between decibels and power changes:

dB Value	Power Increase (Gain)	Power Decrease (Loss)
1 dB	× 1.2	× 0.8 (20.5% loss)
2 dB	× 1.6	× 0.6
3 dB	× 2	1/2
6 dB	× 4	1/4
9 dB	× 8	1/8
10 dB	× 10	1/10 (10% loss)

Components in Series and Parallel Circuits

Electrical circuits can be designed in two fundamental configurations: series and parallel. Knowledge of these configurations is essential to analyze circuit behavior.

Series Circuits and Kirchoff's Voltage Law

In a series circuit, all components are connected along a single path for current flow. Current travels through each component sequentially. The total voltage across the entire series circuit is the sum of the voltage drops across each component. This relationship is described by Kirchhoff's Voltage Law (KVL), which states that the sum of the voltages around a closed loop in a circuit must equal zero. Mathematically, in a series circuit, $V = V_1 + V_2 + V_3$, where:

- V is the total voltage across the series circuit.
- V_1, V_2, V_3 are the voltage drops across each component in the series circuit.

Parallel Circuits and Kirchoff's Current Law

In a parallel circuit, components are connected across each other with two common electrical nodes. The current is divided and flows through each branch independently. The total current supplied by the voltage source in a parallel circuit equals the sum of the currents that flow through each branch. This principle is captured by Kirchhoff's Current Law (KCL), which states that the sum of currents that enter a junction in a circuit must equal the sum of currents that leave the junction. Mathematically, in a parallel circuit: $I = I_1 + I_2 + I_3$, where:

- I is the total current supplied by the voltage source.
- I_1, I_2, I_3 are the currents that flow through each branch in the parallel circuit.

Add Components

Characteristic	Add in Series	Add in Parallel
Resistance (R).	Increases ($R_{total} = R_1 + R_2 + R_3$).	Decreases ($1/R_{total} = 1/R_1 + 1/R_2 + 1/R_3$).
Inductance (L).	Increases ($L_{total} = L_1 + L_2 + L_3$).	Decreases (complex formula for multiple inductors).
Capacitance (C).	Decreases ($1/C_{total} = 1/C_1 + 1/C_2 + 1/C_3$).	Increases ($C_{total} = C_1 + C_2 + C_3$).

Modify Resistor Values

To alter the overall resistance in a circuit, you can use the following techniques:

- **Increase Resistance**: Connect a resistor in series with the existing resistor. The circuit's total resistance will be the sum of the individual resistor values. For example, the addition of two 50-ohm resistors in series to a 120-ohm resistor will give a resistance of 220 ohms.

- **Decrease Resistance**: Add resistors in parallel to the existent resistor. This configuration divides the current flow and produces a lower overall resistance than any single resistor value in the parallel network. When resistors of equal value are connected in parallel, the total resistance will be a fraction of the value of a single resistor. For example, three 90-ohm resistors in parallel would have a combined resistance of 30 ohms.

  Exam Tip: When you calculate the total resistance of resistors in parallel, a quick way to estimate the answer is to choose a value smaller than the smallest resistor in the group but not a value that is extremely small.

Modify Inductor Values

Remember that an inductor's primary opposition to current flow with AC signals stems from its reactance, which functions similarly to resistance.

- **Increase Inductance**: Connect an inductor in series with the existent inductor. The total inductance will increase, and the combined value is the sum of the individual inductors. For example, if a 15-mH inductor is added in series to a 30-mH inductor circuit, the inductance will increase to 45 mH.
- **Decrease Inductance**: Add inductors in parallel. This provides alternative paths for current flow and produces a lower overall inductance than any single inductor in the circuit. For multiple inductors of equal value connected in parallel, the total inductance is calculated similarly to resistors: divide the value of a single inductor by the number of parallel inductors. For example, three 9 mH inductors in parallel would have a combined inductance of approximately 3 mH.

  Exam Tip: To quickly estimate the combined inductance of parallel inductors, select an answer smaller than the smallest inductor value in the group but not a value that is extremely small.

Modify Capacitor Values

Capacitors behave oppositely to resistors and inductors.

- **Decrease Capacitance**: Connect capacitors in series. The total capacitance in a series circuit will be less than the value of any individual capacitor. For instance, when a 10-μF capacitor is connected in series to a 20-μF circuit, the total capacitance reduces to 6.7 μF.
- **Increase Capacitance**: Add capacitors in parallel. This configuration effectively increases the plate area and generates a larger total capacitance. For

example, the addition of two eight nanofarad (nF) capacitors in parallel with a 1,500 picofarad (pF) capacitor will produce a total capacitance of approximately 9.5 nF.

Transformers

How Transformers Work

Transformers are passive electrical devices that transfer electrical energy from one circuit (primary) to another (secondary) by mutual inductance. They consist of two or more coils of wire wrapped around a ferromagnetic core. The primary winding is the coil where electrical power is applied. The secondary winding is the coil where electrical power is extracted.

When AC voltage is applied to the primary winding, it generates a magnetic field within the core. This induces a voltage in the secondary winding per the principle of mutual inductance. This phenomenon works in both directions. That is, voltage applied to the secondary can also induce voltage in the primary.

Reciprocality of Primary and Secondary Windings

Transformers are reciprocal devices. That means that their primary and secondary windings can be interchanged. For example, if the windings in a 5:1 step-down transformer are reversed, it will transform into a 5:1 step-up transformer. Consequently, the secondary voltage will become five times greater than the voltage applied to the new primary winding.

Turns Ratio and Voltage Transformation

The ratio of the number of turns in the secondary winding (N_s) to the number of turns in the primary winding (N_p) is known as the turns ratio. This ratio determines how voltage and current are transformed between the primary and secondary circuits.

- **Step-Down Transformers**: If the secondary winding has fewer turns than the primary ($N_s < N_p$), the transformer acts as a step-down transformer. In this case, the secondary voltage (E_s) will be lower than the primary voltage (E_p) by a factor equal to the turn ratio. For example, a transformer with 100 turns on the primary and 50 turns on the secondary would have a turns ratio of half. This configuration will step down the voltage by a factor of two. The secondary voltage (E_s) will be half the primary voltage (E_p). In this case, if E_p is 200 volts, E_s will be 100 volts.

- **Step-Up Transformers**: If the secondary winding has more turns than the primary ($N_s > N_p$), the transformer acts as a step-up transformer. In this scenario, the secondary voltage (E_s) will be higher than the primary voltage (E_p) by a factor equal to the turns ratio. For example, consider a transformer with a 500-turn primary and a 1,500-turn secondary. The turns ratio here is three (1,500 turns ÷ 500 turns). With 200 volts applied to the primary, the secondary voltage (E_s) would be three times greater and produce an output of 600 volts.

Wire Size and Step-Up Transformers

When transformers are designed, the wire size for the primary and secondary windings must be carefully selected to accommodate the currents that flow within each coil. The relationship between current and wire size is especially critical for transformers that create significant voltage differences between the primary and secondary circuits.

In step-up transformers, where the secondary voltage is higher than the primary voltage, the current in the primary winding will be proportionally higher than in the secondary winding. To handle this higher current, the primary winding is constructed with a larger diameter wire than the secondary winding.

Use of an undersized wire for the primary winding in a step-up transformer poses severe risks. The inadequate wire gauge cannot safely accommodate the higher current that flows through the primary coil. This causes excessive heat generation, which can damage the wire's insulation and increase the likelihood of a short circuit. In extreme cases, the wire may overheat, melt, and cause complete transformer failure. In addition, the increased resistance caused by the undersized wire reduces the transformer's overall efficiency and causes unnecessary power loss.

Chapter 6: Circuit Components

Resistors and Resistance

Resistors are fundamental electronic components that introduce opposition to current flow. Although they share the same basic function, resistors can differ in these key characteristics: nominal resistance value, tolerance, and temperature coefficient.

Resistor Characteristics

Resistor Type	Applications	Advantages	Disadvantages
Carbon Composition.	General use and low-power applications.	It is inexpensive and produces low noise.	It has a lower tolerance (5% or 10%) and a higher temperature coefficient.
Carbon Film.	General use and low-power applications.	It is inexpensive and produces low noise.	It has a lower tolerance (5% or 10%).
Metal Film.	General use. Also has a wider power range.	It produces lower noise than carbon comp/film and has good stability.	It is more expensive than carbon comp/film.
Wire-wound.	High power applications.	It can handle high wattage.	It has high inductance (unsuitable for RF circuits) and is larger than the other types.

Metal Oxide.	High power applications. It is non-inductive.	It has high power handling, is non-inductive for RF, and has good stability.	It is more expensive than other types.

Nominal Resistance Value

The nominal resistance value is the intended resistance of a resistor. Common nominal values range from 1 Ω to over 1 MΩ. It is the main value marked in ohms (Ω) or through color bands on the resistor. The nominal value guides your choice of resistor to achieve the desired current limitation level within a circuit. For instance, resistors might be used to limit current flow through LEDs on a control panel, set bias voltage for amplifier stages, or adjust gain levels in signal processing circuits.

Resistor Tolerance

Resistor tolerance indicates the acceptable range of deviation from the nominal resistance value and is usually expressed as a percentage. A 220 Ω resistor with a 10% tolerance could have a resistance anywhere between 198 Ω and 242 Ω. Knowledge of the tolerance range is important in circuit design. High-precision circuits like filter networks or crystal oscillators might require resistors with tighter tolerances (1% or less) to ensure accurate signal behavior.

Temperature Coefficient

This parameter describes how a resistor's value changes with temperature fluctuations. Knowledge of the temperature coefficient is important for ham radio operators because many radio components operate in environments where temperature variations can occur. For instance, a transmitter might generate heat during operation. Resistors with low-temperature coefficients help maintain circuit performance because they minimize resistance changes caused by temperature swings. This is important for circuits where consistent resistance values are vital for signal stability or proper biasing.

Parasitic Inductance in Wire-Wound Resistors

Parasitic inductance refers to an undesirable inductive characteristic that arises unintentionally due to the physical construction of certain electronic components. In the case of wire-wound resistors, the coiled wire and form on which it is wound create a small amount of inductance.

Parasitic inductance can have negative consequences for RF circuits. The inductance can interact with other circuit elements in unintended ways and cause unpredictable circuit behavior, altered frequency response, and system instability. To minimize parasitic inductance in RF circuits, resistors with very low inductance such as carbon composition, metal film, and metal oxide resistors are used.

Inductors and Inductance

Inductors are passive components that store energy in a magnetic field when current flows through them. They are used in various circuits and differ based on core material, construction, and adjustability.

Inductors: Types and Applications

- **Laminated Iron Core**: This is commonly used for DC and AC power supply filters.
- **Powdered Iron Solenoid**: This is found in RF chokes, power supplies, and low- to medium-frequency circuits.
- **Powdered Iron & Ferrite Toroids**: This is widely used in audio and radio circuits.
- **Air Core**: This is mostly designed for high-frequency RF transmissions.

Variable Inductors

Variable inductors allow the adjustment of inductance. This makes them useful in tuning circuits.

- **Low-Power**: To adjust the inductance, a magnetic core moves within the coil.
- **High-Power**: Inductance is adjusted via a sliding contact (like roller inductors found in antenna-matching networks).

Fundamental Principles of Inductance

Inductance is the inherent property of an inductor to store energy in a magnetic field. Inductance is measured in units called henries (H). The amount of inductance is directly influenced by the permeability of the inductor's core. A core with higher permeability enables greater storage of magnetic energy and produces increased inductance. The core material must be carefully selected as it determines the frequency range for which the inductor is best suited.

Coupling Between Inductors

When inductors are situated close to each other, their magnetic fields can interact to produce the phenomenon known as coupling or mutual inductance. This mechanism allows signal transfer between circuits without a direct electrical connection. Unintended coupling can be minimized when solenoidal inductors are positioned at right angles to one another.

Ferrite Core Toroidal Inductors

Ferrite core toroidal inductors provide a range of benefits. They can achieve high inductance values. Their magnetic properties can also be tailored to specific frequency ranges. Furthermore, their design helps to contain the magnetic field effectively. The key to optimal performance requires the right ferrite core composition or "mix" to align with the inductor's intended operating frequency range.

Capacitors and Capacitance

Capacitor Fundamentals

Capacitors are essential components in electronics and consist of two conductive surfaces (electrodes) separated by insulation material known as a dielectric. This simple construction allows a capacitor to store electrical energy within the electric field created between the electrodes. Capacitance, measured in farads (F), represents a capacitor's ability to store electrical charge. An important characteristic of capacitors is their ability to block the flow of DC but allow AC to pass.

Capacitor Types: Applications and Advantages

Capacitors come in various forms, each with characteristics suited to different applications.

- **Ceramic Capacitors**: Ideal for RF filters and bypasses at high frequencies. Ceramic capacitors are known for their relatively low cost.

- **Electrolytic Capacitors**: Commonly used in power supply filter circuits. Their key advantage is their provision of high capacitance values within a compact size.

- **Plastic Film Capacitors**: Well-suited for audio circuits and lower radio frequencies.

- **Silvered-mica Capacitors**: They are highly stable, known for low losses, and are suitable for RF circuits.

- **Air and Vacuum Dielectric Capacitors**: Designed for transmissions and high-power RF circuits.

Capacitor Polarity

Aluminum and tantalum electrolytic capacitors use a thin oxide film as their dielectric layer. This construction allows them to achieve remarkably high capacitance values within a relatively small physical size. However, they are polarity sensitive. That is, the correct-voltage polarity must be strictly observed. Incorrect polarity can cause short circuits, damage the dielectric layer, and even cause the capacitor to overheat and explode.

Reactance and Impedance

Opposition to the Flow of AC

In AC circuits, the elements that oppose current flow extend beyond simple resistance. The table summarizes the key factors that influence impedance (Z)—the total opposition to AC flow—in electrical circuits. Unlike DC circuits, where R solely dictates opposition, AC circuits involve a more complex phenomenon.

Type of Opposition	Symbol	Unit	Varies with Frequency?	Notes
Resistance.	R.	Ohms (Ω).	No.	The same resistance value is found in DC circuits.
Reactance.	X.	Ohms (Ω).	Yes.	It varies based on the frequency and type of reactance—capacitive (X_C) or inductive (X_L).
Impedance.	Z.	Ohms (Ω).	Yes.	The combination of resistance and reactance. Its value depends on frequency.

Reactance

Unlike simple resistance, reactance refers to the opposition to AC flow caused by the presence of capacitors or inductors in the circuit. It is denoted by the symbol "X" and, like resistance, is measured in ohms (Ω).

Types of Reactance

- **Capacitive Reactance (X_C)**: Capacitors oppose changes in voltage. The degree of this opposition, known as capacitive reactance, is inversely proportional to the frequency of the AC signal. As frequency increases, capacitive reactance (X_C) decreases. This relationship is expressed as $X_C = 1 \div (2\pi fC)$, where f is frequency and C is capacitance.

- **Inductive Reactance (X_L)**: Inductors resist changes in current. The opposition created is called inductive reactance and is directly proportional to frequency. As frequency increases, inductive reactance (Xl) increases. The formula to calculate inductive reactance is $X_L = 2\pi fL$, where f is frequency and L is inductance.

Resonance in AC Circuits

Resonance is a unique phenomenon in AC circuits that contain both inductance (L) and capacitance (C). At a specific frequency, known as the resonant frequency, the inductive reactance (X_L) and capacitive reactance (X_C) become equal and opposite, so they effectively cancel each other out.

Effects of Resonance in Series LC Circuits

In a series LC circuit at resonance, the inductive and capacitive reactances nullify each other. This leaves only the circuit's resistance to oppose current flow and produces a very high overall current within the circuit. Thus, the circuit effectively behaves like a short circuit at the resonant frequency.

Applications of Resonance

- **Filters**: Resonant circuits are used to create filters that either pass or block specific frequencies, which enables signal selection.

- **Tuned Circuits**: Resonant circuits are integral to tuning devices like radios. They help specific frequencies from a broad range of signals.

Parasitic Capacitance in Inductors

Even real-world inductors possess a small amount of inherent capacitance, referred to as parasitic capacitance. This capacitance arises due to the small spaces between adjacent turns of wire in the coil. Although usually very small, its effect becomes more noticeable at higher frequencies. Capacitive reactance decreases as frequency increases because capacitors oppose changes in voltage, and higher frequencies signify more rapid voltage changes. On the other hand, inductive reactance increases as frequency increases because inductors resist changes in current, and higher frequencies correspond to faster current fluctuations.

Self-Resonance

Due to the effects described above, a specific frequency exists at which the inductive reactance (X_L) and capacitive reactance (X_C) within an inductor become equal. This phenomenon is known as self-resonance. Above the self-resonant frequency, the inductor's overall behavior becomes predominantly capacitive.

Impedance

Impedance, measured in Ω, represents the total opposition to the flow of AC within a circuit. Impedance (Z) stems from the combined effects of two key factors: resistance (R) and reactance (X).

Mathematical Representation of Impedance

Ohm's Law, which applies to both DC and AC circuits, states that the voltage (V) across a component is equal to the current (I) that flows through it multiplied by its impedance (Z): $V = IZ$. We can rearrange this formula and make impedance (Z) the subject to get $Z = V \div I$. Therefore, the impedance in an AC circuit is the ratio of the voltage and the current in that circuit.

The Inverse of Impedance: Admittance

The inverse of impedance is admittance (Y). If the impedance is the opposition to current flow across an AC circuit, admittance represents the ease with which current flows through a circuit. Admittance is calculated as $Y = 1 \div Z$ and is the ratio of current to voltage in a circuit (that is, $Y = I \div V$). Similar to how higher impedance signifies greater opposition to current, higher admittance indicates a more conductive path for current flow.

How to Match Impedance Maximum Power Transfer

Maximum power transfer from a source (such as a radio transmitter) to a load (such as an antenna) is essential for optimal signal transmission. This is where impedance matching comes into play. The principle behind impedance matching is that maximum power transfer occurs when the source impedance (Z_S) and the load impedance (Z_L) are equal.

Impedance in Amateur Radio

Amateur radio transmitting equipment is usually designed with a standard output impedance of 50 Ω. Ideally, antennas should also present a feed point impedance of 50 Ω for direct connection. However, in reality, an antenna's impedance can fluctuate across different frequencies, which makes it difficult to maintain a perfect match throughout its operational range.

Techniques to Match Impedance

As discussed below, several techniques can be used to match impedance in AC circuits.

- **Impedance-Matching Transformers:** A transformer can be used to bridge a mismatch. Its function transforms the impedance seen at the antenna's feed point to the desired 50 Ω impedance required by the transmitter and feed line. Technically, a transformer tricks the feed line and antenna, so they think they have an impedance of 50 Ω, regardless of the actual impedance. This ensures efficient power transfer and prevents power from being reflected back into the transmitter, which could cause damage. The turns ratio of a transformer is central to its capacity to match impedance. The turns ratio is calculated as the square root of the impedance ratio. The impedance ratio is the ratio of the feed line impedance and the transmitter impedance.

  Mathematically, *Turns Ratio (TR) = √(Impedance Ratio)*, where *Impedance Ratio = Feed Line Impedance ÷ Transmitter Impedance.*

- **LC Pi Networks**: These consist of specific arrangements of inductors (L) and capacitors (C) to create a desired circuit. Most circuits used to match impedance are inductors and capacitors.

- **Coaxial Cable Length**: The length of a coaxial cable can be strategically chosen to introduce a specific amount of reactance, which can help match impedance.

Calculation Examples

Example: What is the turns ratio for a transformer that matches an 800-ohm antenna feed point impedance to that of a 50-ohm coaxial cable?

Solution:

- Applicable Formula:

  Turns Ratio (TR) = √(Impedance Ratio), where Impedance Ratio = Antenna Feed Point Impedance ÷ Coaxial Cable Impedance.

- Given Information:

  Feed Line Impedance = 800 ohms.

  Transmitter Impedance = 50 ohms.

- Steps:

  First, calculate Impedance Ratio = (800 ohms) ÷ (50 ohms)= 16.

  Then, find the square root to get the Turns Ratio: √16 = 4.

Rectifiers

Rectifiers are circuit components that convert AC into DC. Two main rectifier circuit types are important for radio circuits: half-wave and full-wave.

- **Half-Wave Rectifier**: A half-wave rectifier is a circuit that permits electrical current to flow only during one-half (180 degrees) of the input AC waveform. This means that the output consists of a pulsating DC voltage suitable for simple, low-power applications. Half-wave rectifiers are simple and require just one diode for implementation.

- **Full Wave Rectifier**: Full-wave rectifiers provide more efficient AC to DC conversion because they use both halves of the AC input waveform. Two common types of full-wave rectifier circuits are discussed below.

  - **Full-Wave Bridge Rectifier:** This configuration uses four diodes arranged in a bridge formation. A bridge rectifier produces a DC output across the entire 360-degree cycle of the AC input waveform. The output of a full-wave bridge rectifier consists of a series of DC pulses. The frequency of these output pulses is double the frequency of the input AC voltage. This higher pulse frequency simplifies the filtration process required to produce a smoother DC output.

- **Center-Tapped Transformer Rectifier:** This type of rectifier employs a transformer with a center-tapped secondary winding and two diodes. The center tap of the transformer acts as a reference point, with each diode conducted on alternate halves of the input AC waveform.

Active Components

Semiconductor Components

Diodes

Diodes are fundamental semiconductor components built from different materials:

- **P-type Material**: This contains an abundance of positive charges (holes).
- **N-type Material**: This contains an excess of negative charges (electrons).

The PN Junction

The interface between the P-type and N-type materials forms a PN junction. This junction is critical for the diode's core function: a barrier that allows current to flow in only one direction.

- **Forward Bias**: When a positive voltage is applied to the P-type material (anode) and a negative voltage to the N-type material (cathode), the diode is forward-biased and allows current to flow. When a forward bias voltage is applied to the diode, the positive voltage at the anode (P-type) attracts holes in the P-type region, and the negative voltage at the cathode (N-type) attracts electrons. This combined effect pushes more electrons toward the junction. These additional electrons have enough energy, provided by the external voltage, to overcome and cross the barrier. The flow of these majority carriers (electrons) constitutes the forward current through the diode. The forward voltage drop (around 0.3 V for germanium and 0.7 V for silicon) represents the minimum external voltage required to provide enough energy for sufficient electrons to overcome the barrier and establish a noticeable current flow.

- **Reverse Bias**: When the polarity is reversed, the diode blocks current flow.

Diode Ratings

Diode ratings are critical parameters that define the operational limits of diodes. If these ratings are exceeded, it can cause catastrophic diode failure due to breakdown or overheating, which could potentially compromise the entire circuit. These ratings, such as the peak inverse voltage (PIV), average forward current, and junction capacitance, are essential for safe, reliable, and efficient performance.

- **Peak Inverse Voltage (PIV)**: PIV is the maximum voltage a diode can withstand in the reverse-bias condition (when voltage polarity is applied to block current flow) before it experiences a destructive breakdown. This breakdown occurs when the electric field stresses the semiconductor material and causes an uncontrolled surge of reverse current. If you exceed the PIV, this can permanently damage the diode and other components in the circuit.

- **Average Forward Current**: This rating specifies the maximum continuous forward current a diode can safely handle before excessive heat buildup occurs. When current flows through a diode, even in the forward direction, energy is lost as heat due to internal resistance. If the current exceeds the average forward current rating, the heat generated can cause irreversible damage to the delicate structure of the diode.

- **Junction Capacitance**: In reverse-bias mode, the depletion region at the junction of a diode behaves like a small capacitor. The junction capacitance measures this parasitic capacitance, which becomes especially significant in high-frequency circuits. A larger junction capacitance makes it more difficult for the diode to switch rapidly between conducting and non-conducting states. For high-speed switching applications, diodes with low junction capacitance are preferable.

Specialized Diode Types

The following diode types have applications in RF devices:

- **PIN Diode**: These diodes have a layer of intrinsic semiconductor material between the P and N regions. This increases the diode's depletion region, making them suitable for RF switching and attenuation.

- **Schottky Diode**: Schottky diodes are characterized by a metal-semiconductor junction rather than a PN junction. This produces minimal junction capacitance and enables operation at very high frequencies.

- **Varactor Diode**: A varactor diode acts as a voltage-controlled capacitor when reverse-biased. This property is used in tuning circuits.

- **Zener Diode**: Zener diodes are designed to operate reliably in the reverse breakdown region and maintain a relatively steady voltage. They are commonly used for voltage regulation purposes.

Transistors

Transistors are fundamental semiconductor devices that exhibit remarkable amplification properties. This characteristic makes them exceptionally useful as switches that help control voltage and current flows in electronic circuits. They operate either in saturation or cutoff modes, which makes them effectively function as electronic switches for voltage and current.

Operating Modes in Switching Applications

In a bipolar transistor configured as a switch, saturation, and cutoff represent the two primary operating states.

- **Saturation (On)**: In saturation mode, a transistor allows maximum current flow and resembles a closed switch. Any further increase in the input signal does not produce an increase in output current.

- **Cutoff (Off)**: Conversely, the transistor effectively blocks current flow and acts like an open switch in the cutoff mode.

Transistors in Logic Circuits

The saturation and cutoff modes of transistors map directly to digital logic states. A transistor in saturation corresponds to a logic "1" (high), while a transistor in cutoff represents a logic "0" (low). This forms the basis for transistors used to construct logic gates and digital circuits.

Bipolar Junction Transistors (BJTs)

Bipolar junction transistors (BJTs) are constructed with three semiconductor layers and corresponding electrodes: Collector (C), Base (B), and Emitter (E).

BJTs are current-controlled devices. This means a small current that flows between the base and emitter controls a larger current flow between the collector and emitter. This principle is analogous to how a valve can control water flow through a fire hose. A small input (such as a valve) controls the flow of a much larger volume of water. The ratio of the larger collector-emitter current to the smaller base-emitter current is known as the current gain or beta (β).

Field Effect Transistors (FETs) and MOSFETs

FETs also have three electrodes: Drain (D), Source (S), and Gate (G). The drain is the output region, the source is the reference or ground region, and the gate is the control terminal. Unlike BJTs, FETs are voltage-controlled devices. The voltage applied across the gate, and source terminals regulates the current flow between the drain and source.

An FET type known as a MOSFET (metal oxide semiconductor field effect transistor) has a thin insulating oxide layer on the gate. This insulation gives MOSFETs certain advantages in power efficiency and switching speed.

Vacuum Tubes

Vacuum tubes, also known as electron tubes, have been largely replaced by modern transistors. Although less prevalent today, they still find use in specialized applications. Vacuum tubes control the flow of electrons within an evacuated glass enclosure. Amplified vacuum tubes usually contain at least three electrodes.

Important Safety Note: Vacuum tubes usually operate with extremely high voltages (often in the range of 2,000 to 3,000 volts). Caution must be exercised with equipment that contains vacuum tubes to avoid severe electrical shock.

Key Elements of Vacuum Tubes

- **Filament or Heater**: This element heats the cathode to emit electrons through a process called thermionic emission.
- **Cathode**: The cathode serves as the primary source of electrons.
- **Plate (Anode)**: The plate is the element that collects the electrons that flow through the vacuum tube.
- **Control Grid**: The grid regulates the electron flow from the cathode to the plate. Small variations in the voltage applied to the control grid significantly impact the electron flow and can be used for amplification.
- **Screen Grid**: Found in tetrode and pentode tubes, the screen grid mainly reduces the capacitance between the control grid and the plate. This reduction in capacitance enhances the tube's stability and performance, especially at higher frequencies.

Analog and Digital Integrated Circuits

Analog Integrated Circuits

Analog ICs, often called chips, are designed to process continuous signals that can vary in amplitude over a range of values. Their main applications include:

- Amplification: This boosts the strength of electrical signals.
- Filters: This involves selective attenuation or enhancement of specific frequencies.
- Measurement: This involves analysis and quantification of physical parameters.
- Power Control: This involves the regulation and management of power delivery to circuits.

Examples of common analog ICs include linear voltage regulators and operational amplifiers. Linear voltage regulators maintain a stable output voltage even as the load current varies. Operational amplifiers (op-amps) are versatile building blocks in many analog DC and audio circuits.

RF Integrated Circuits

RF ICs are specialized integrated circuits designed to handle signals within the range of radio frequencies. These high-frequency ICs are tailored for functions such as:

- **Amplification**: This increases the signal strength of RF signals.
- **Mixing**: This combines or separates signals of different frequencies.
- **Modulation/Demodulation**: This encodes or decodes information in radio frequency signals.
- **Filtering**: This isolates desired signal bands within the RF spectrum.

MMICs (monolithic microwave integrated circuits) are a special type of RF-integrated circuit that operate at even higher frequencies (which extend into microwave ranges).

Digital Integrated Circuits

Digital ICs operate on discrete voltage and current levels that represent the binary values of 0 and 1. The binary states of 1 and 0 can be readily represented by "on" and "off" conditions in electronic circuits. Digital circuits can perform various functions like computation (such as mathematical operations and logical algorithms) and control (which governs the operation of devices and systems). CMOS (complementary metal oxide semiconductor) technology is widely employed in digital ICs due to its high speed and low power consumption.

Digital Logic Basics

The Gates

Digital electronic circuits are constructed from fundamental building blocks known as logic gates. These gates perform basic logical operations on binary signals (0 or 1), the language of digital systems. Two of the most common logic functions are:

- **AND**: The output of an AND gate is high (1) only if all its inputs are high (1).
- **OR**: The output of an OR gate is high (1) if at least one of its inputs is high (1).

Truth Tables

Truth tables provide a concise way to represent the behavior of logic gates. They list all possible input combinations and outputs. Below are the truth tables for the AND and the OR gates.

The AND Gate: The output of the AND gate is high (1) only when both inputs A and B are high (1).

Input A	Input B	Output
0	0	0
0	1	0
1	0	0
1	1	1

The OR Gate: The output of the OR gate is high (1) if at least one of its inputs, A or B, is high (1).

Input A	Input B	Output
0	0	0
0	1	1
1	0	1
1	1	1

Inversion

Inversion is another fundamental operation in digital logic, where a signal is flipped from its original state (0 to 1 or 1 to 0). This function is achieved with an inverter called a NOT gate. Inverted forms of other gates, like NAND (NOT-AND), are also common in digital design.

The NOT Gate: The NOT gate inverts the input signal. This means:

- If the input is 0, the output will be 1.
- If the input is 1, the output will be 0.

Input A	Output
0	1
1	0

Combined Gates and Sequential Logic

Digital circuits often go beyond simple combinations of logic gates to incorporate memory elements and sequential behaviors. This is accomplished through flip-flops and the interconnected arrangements that form the foundation of more complex digital systems.

- **Flip-Flops**: A flip-flop is a basic digital circuit with two stable states, usually represented by a binary 0 or 1. This ability to "remember" a previous state makes flip-flops essential for sequential logic circuits, where the output depends on both the current inputs and the circuit's prior state.
- **Counters**: Various sequential circuits can be designed when flip-flops are connected and their outputs are fed back as inputs. A prime example is a digital counter. A three-bit counter, constructed with three flip-flops, can count through $2^3 = 8$ distinct states (0 to 7). Similarly, a four-bit counter can represent $2^4 = 16$ different states.
- **Shift Registers**: Shift registers are sequential circuits that shift data along a chain of storage elements (such as flip-flops). Each clock cycle causes the data

bits to move one position within the shift register. Shift registers have applications in data transfer, storage, and serial-to-parallel conversion operations.

Microprocessors and Components

A microprocessor is a complex integrated circuit (IC). It is a computer compacted within a single chip. Essentially, it is a miniature digital processing unit constructed from millions of microscopic transistors arranged as digital logic gates. Microprocessors can execute vast numbers of instructions every second. Nearly all microprocessors are based on the CMOS logic.

Key Components of Microprocessors

Besides the central processing unit itself, microprocessors usually integrate features such as:

- **Input/Output (I/O) Ports**: These interface for communication with external devices and peripherals. These can be parallel (they handle multiple bits simultaneously) or serial (they transmit one bit at a time).
- **Counters and Timers**: These elements keep precise times, schedules, and generate specific time intervals for various operations.
- **Memory Interface**: Microprocessors usually rely on external memory chips to store programs and data due to space limitations on the chip itself.

Visual Displays

Visual displays present information from electronic devices in a human-readable format. These displays can range from simple indicators to advanced screens that display complex text or graphics. The two common display technologies are light-emitting diodes (LEDs) and liquid crystal displays (LCD).

Light-Emitting Diodes (LEDs)

LED display technology is built on semiconductor devices. LEDs emit visible light when a forward-bias voltage is applied to their PN junctions. This property makes them useful as indicators and visual displays. They come in various colors and are known for their energy efficiency, quick response times, and long life span.

Liquid Crystal Displays (LCDs)

LCDs operate based on a unique property of liquid crystals. This specialized material is sandwiched between transparent panels. Through the application of precise voltages to electrodes embedded within the panels, the alignment of the liquid crystal molecules can be manipulated. This controlled alignment determines whether light is blocked or allowed to pass through specific display sections.

One fundamental property is that the liquid crystal layer within an LCD does not emit light on its own. Therefore, LCDs require an external light source for visibility. This source can be ambient light in the environment or dedicated backlights integrated into the display. LCDs are a good choice for devices often used outdoors or in areas with strong ambient lights.

Batteries and Discharge

Large storage batteries, such as those found in marine or RV applications, can be used to provide emergency backup power due to their ability to deliver sustained power over extended periods. A battery should not be discharged below its minimum allowable voltage. Over-discharge reduces a battery's life span and overall capacity. For lead-acid batteries, this minimum voltage is usually around 11.8-12.0 V for a full discharge, although discharging below 10.5 V can cause permanent damage.

Some Battery Types and Applications

Batteries with low internal resistance, such as NiCad (nickel-cadmium) and lead-acid types, can deliver high discharge currents. This characteristic can be advantageous in applications that require brief bursts of high power.

Lithium iron phosphate (LiFePO4) batteries are popular for portable operations due to their favorable energy density, life span, and safety characteristics.

Batteries and Self-Discharge

All batteries undergo self-discharge, a gradual loss of stored energy over time. Store batteries in a cool, dry environment to minimize this effect and preserve battery capacity. Extremely cold temperatures can cause irreparable damage to batteries that contain liquid or gel electrolytes.

Chapter 7: Practical Circuits

AC Power

In DC circuits, electrons flow in a single, constant direction. Imagine water that flows steadily through a pipe. DC is usually obtained from batteries, solar cells, and certain electronic components. The voltage and current in a DC circuit remain constant over time, often depicted as a straight line on a graph.

In AC circuits, the direction of electron flow periodically reverses. This back-and-forth motion is similar to waves in the ocean. AC is the electricity delivered to our homes and powers most appliances. AC voltage and current constantly change direction and are usually represented as sinusoidal waves on a graph.

Alternating Current

AC power is more complex than DC power due to voltage and current flow that constantly reverse direction. This reversal makes it necessary to use a different measurement to compare AC to DC effectively.

Root Mean Square (RMS)

The RMS value of an AC voltage is defined as the equivalent DC voltage that would produce the same heat effect (power dissipation) within a resistor. The RMS value is used to express the effective voltage or current of an AC waveform. It allows us to make meaningful comparisons between AC and DC power. For instance, the standard 120-volt AC power outlet has an RMS voltage of 120 volts. This value means the outlet delivers the same effective heating effect in a resistor as a 120-volt DC source. To calculate the RMS value:

- Square each instantaneous voltage value over one cycle of the waveform.
- Find the average (mean) of these squared values.
- Take the square root of the average.

Relationship Between RMS, Peak, and Peak-to-Peak Values

In sinusoidal AC waveforms, three important voltage measurements help describe waveform characteristics:

- **Root Mean Square (RMS) Voltage**: This is the effective voltage of an AC waveform. It represents the equivalent DC voltage that would produce the same amount of power dissipation in a resistive load. The RMS voltage is 0.707 times the peak voltage of the AC waveform. RMS = V_{pk} × 0.707.

- **Peak Voltage (Vpk)**: This is the maximum positive or negative voltage reached during one cycle of the AC waveform. To obtain peak voltage from the RMS voltage, multiply the RMS voltage by 1.414. Mathematically, Vpk = Vrms × 1.414.
- **Peak-to-Peak Voltage (Vp-p)**: This is the total voltage difference between the positive and negative peaks of the waveform. Peak-to-peak voltage is the peak voltage multiplied by 2. Mathematically, Vp-p = Vpk × 2. Alternatively, Vpk = Vp-p ÷ 2.

RMS and Peak Voltage Calculation Examples

Example 1: What is the RMS voltage of a sine wave with a value of 25 volts peak?

- RMS = Vpk × 0.707.
- Here, Vpk = 25 V.
- Substitute for the variable: RMS = 25 V × 0.707 = 17.675 V.
- RMS ≃ 17.68 V.

Example 2: Determine the peak-to-peak voltage of a sine wave with an RMS voltage of 100.0 volts.

- Vpk = RMS × 1.414.
- Here, RMS = 100.0 V.
- Substitute for the variable: Vpk = 100.0 V × 1.414 = 141.4 V.
- To get the peak-to-peak voltage (Vp-p), substitute for Vpk in the formula Vp-p = Vpk × 2.
- Vp-p = 141.4 V × 2 = 282.8 V.
- Vp-p = 282.8 V

Example 3: What is the RMS voltage across a 75-ohm dummy load that dissipates 800 watts of power?

- Required value: Voltage (E).

- Given information: Power = 800 watts (W) and Resistance = 75 ohms (Ω).
- Applicable formula: E (Volts) = $\sqrt{(P \times R)}$.
- Substitute applicable values: P = $\sqrt{(800 \times 75)} = \sqrt{60,000} = 244.9490$ V.

Peak Envelope Power (PEP)

Definition and Measurement

PEP is a standard way to measure the maximum output power of a radio transmitter, especially for SSB voice transmissions. In amateur radio, the FCC limits the maximum PEP output to 1500 watts. PEP measurement is essential because it directly indicates the maximum power output of a transmitter, which influences communication range and signal strength.

PEP is defined as the average power supplied to the antenna during one complete RF cycle at the peak of the modulation envelope. Modulation envelope refers to the shape of the signal that represents your voice information superimposed on the RF carrier wave. Peak envelope voltage (PEV) is the related term that represents the maximum voltage reached during that peak of the signal's envelope.

PEP and Average Power

The relationship between PEP and average power depends on the modulation type.

- Constant Power Modes: For unmodulated signals (key-down CW, unmodulated AM carrier, and FM), PEP is equal to average power.
- Varied Power Modes: In modes like SSB, where the power varies within the modulation envelope, PEP will be higher than the average power.

For example, an unmodulated carrier with an average power measurement of 1,060 watts on a wattmeter will have a PEP of 1,060 watts. The PEP to average power ratio in this case is 1.0.

PEP Calculation Examples

Example 1: A transmitter is connected to a 50-ohm dummy load. An oscilloscope measurement across this load indicates a peak-to-peak voltage of 120 volts. Assume a constant-power transmission mode and calculate the output PEP of the transmitter.

- Required Value: PEP in watts (W).

- Given Information:

  Peak-to-peak Voltage (V_{p-p}) = 120 volts.

  Resistance (R) = 50 ohms.

- Applicable Formulas: $P = (E^2 \div R) = (V_{rms}^2 \div R)$. For this formula, E has to be in RMS voltage.

- Solution Steps:

  Convert Peak-to-Peak to Peak Voltage: $V_p = V_{p-p} \div 2 = 120\ V \div 2 = 60\ V$.

  Convert Peak Voltage to RMS Voltage: $V_{rms} = V_p \times 0.707 = 60\ V \times 0.707 \approx 42.4\ V$.

  Calculate PEP with RMS Voltage (E): $PEP = V_{rms}^2 \div R = (42.4\ V)^2 \div 50\ \Omega \approx 36\ W$.

Example 2: A transmitter is connected to a 75-ohm dummy load. If the RMS voltage measured across the load is 125 volts, calculate the output PEP of the transmitter. Assume a constant-power transmission mode.

- Required Value: PEP in watts (W).

- Given Information:

  RMS Voltage (V_{rms}) = 125 volts.

  Resistance (R) = 75 ohms.

- Applicable Formulas: $P = (E^2 \div R) = (V_{rms}^2 \div R)$.

- Solution Steps: $PEP = V_{rms}^2 \div R = (125\ V)^2 \div 75\ \Omega \approx 208.3\ W$.

Filter Circuits

The output of a rectifier, although technically DC, contains a significant amount of ripples—rapid fluctuations in voltage or current. These fluctuations make the output unsuitable to directly power most electronic circuits. Filter networks are used to smooth out this ripple and produce a more constant DC output.

Filter Networks

Filter networks mainly consist of capacitors and inductors arranged in specific configurations to smooth out the ripple variations. Capacitors store energy during voltage peaks and release it during voltage dips to smooth voltage fluctuations. Inductors, on the other hand, oppose changes in current, which smooths out current variations.

Different filter designs can smooth either voltage or current ripple. One approach to smooth the power supply ripple is to use a large electrolytic capacitor. The capacitor charges during the peaks of the rectified waveform and discharges during the troughs. This effectively reduces voltage ripple and provides a more stable DC output.

High Voltage Power Safety

Filter networks can store potentially dangerous high voltages. Safety nets are usually incorporated to prevent lethal electric shocks. Two important measures are discussed below.

- **Primary Fuses:** Fuses installed in the primary circuit of the power supply protect against short circuits and abnormally high current surges that could cause damage or fire.
- **Bleeder Resistors:** Power supplies often incorporate bleeder resistors connected across the filter capacitors. These resistors safely discharge the stored energy within the capacitors after the power supply is turned off to reduce the risk of electric shock.

Caution: Even after a high-voltage power supply is unplugged, you should always wait enough time for the bleeder resistors to discharge the capacitors fully. Never assume the capacitors are safe to handle simply because the device is switched off or disconnected from the main power.

Components of Radio Systems

Radio systems rely on several fundamental components to generate, manipulate, and transmit signals across the electromagnetic spectrum. This section examines how precisely controlled frequencies are generated, how signals are shaped and filtered, and the various ways information is encoded onto these frequencies through modulation.

Oscillators

Oscillators are the foundation of radio systems. They generate stable RF signals that act as carrier waves to carry information (like voice and data). Their output signals must be as close to a pure sine wave as possible for reliable communication. Stable oscillators ensure signals stay within their designated frequency bands to minimize interference with other radio users. The main components common to all oscillator circuits include the following:

- **Amplifier** to boost the signal strength.
- **Positive feedback loop**: A portion of the amplified signal is fed back to the input. When the feedback reinforces the input signal at a specific frequency, sustained oscillation occurs.
- **Frequency-selective filter**: This ensures that the feedback reinforces only the desired frequency to produce sustained oscillation.

Oscillators can be broadly grouped as fixed-frequency oscillators and variable-frequency oscillators.

- **Fixed-frequency oscillators** include LC oscillators and crystal oscillators. LC oscillators use a "tank circuit" with an inductor (L) and capacitor (C) to determine the resonant frequency. However, the inductance or capacitance is usually fixed in LC oscillators, and tuning is achieved by other means in variable-frequency designs. These oscillators are less common in modern equipment as they are less precise than crystal oscillators. Crystal oscillators use the piezoelectric properties of a quartz crystal to provide exceptional frequency stability. They are essential when precise frequency control is required.
- **Variable frequency oscillators (VFOs)** allow flexibility so a radio can be tuned across a range of frequencies. There are different types, such as LC circuits, phase-locked loops (PLL), and direct digital synthesizers (DDS). In LC VFOs, the inductance or capacitance is adjusted in order to vary the frequency. A PLL uses a feedback loop to lock the oscillator to a stable reference frequency. DDS, popular in modern SDRs, uses digital controls for precision, rapid frequency changes, and excellent stability.

Heterodyning

Heterodyning is a signal processing technique pioneered by Reginald Fessenden in 1902. It involves a combination of two frequencies to generate new frequencies. This process is central to many radio communication functions.

Frequency Mixers

Mixers are electronic circuits that perform heterodyning. They allow RF signals to be shifted up or down to different parts of the spectrum. Mixers accept two input frequencies (f1 and f2) and produce four output signals: the two input frequencies (f1 and f2), the sum (f1 + f2), and the difference (f1 – f2) of these frequencies. This is called mixer math.

This conversion allows radios to transmit and receive signals at practical frequencies that match antenna requirements and their assigned operating bands. It also simplifies receiver design and allows most filtration and amplification to occur at a fixed intermediate frequency, which drastically improves performance. In mixers, the RF signal is usually one input (f1), while a locally generated signal from an oscillator (LO) within the receiver radio serves as the other (f2).

Frequency Multipliers

Frequency multipliers are closely related to mixers but serve a slightly different function. Unlike mixers, a multiplier accepts a single input signal and generates a multiple of it. This multiple is called a harmonic. For instance, if the multiplier input is 10 MHz, and it is designed as a doubler (x2), the output will be a clean 20 MHz harmonic signal. Multipliers can be used to generate stable high-frequency signals in FM radio systems. They allow the creation of stable VHF or UHF signals with a lower-frequency oscillator that is multiplied upward.

RF Filters

Radio filters selectively pass the desired frequency ranges and suppress unwanted signals, interference, and noise. This allows receivers to tune into specific channels and improves signal quality for both receivers and transmitters. RF filters are categorized based on their frequency response characteristics:

- **Low-pass filters** allow frequencies below a defined cutoff frequency to pass with minimal attenuation. Signals above the cutoff frequency are progressively attenuated. The cutoff frequency is often defined as the point where the output signal power drops to half (-3dB) of the input power. Hence, the cutoff frequency is sometimes called the "half power point."

- **High-pass filters** permit frequencies above a specified cutoff frequency to pass and attenuate frequencies below this point.

- **Band-pass filters** are designed to pass a specific range of frequencies located between an upper and a lower cutoff frequency. This frequency range is referred to as the filter bandwidth. Signals outside of this defined bandwidth are attenuated.

- **Band-stop filters** attenuate signals within a specific frequency range between upper and lower cutoff points and allow signals outside that band to pass.

- **Notch filters** are a specialized type of band-stop filter with an extremely narrow stopband designed to eliminate a very specific interference frequency.

Key Filter Parameters

- **Insertion loss** means the signal weakens as it passes through the filter, even within the filter's passband. An ideal filter minimizes insertion loss.

- **Ultimate rejection** describes how effectively a filter attenuates signals outside its intended passband. A high ultimate rejection is usually desirable for strong suppression of unwanted signals.

Modulators

Modulators encode information onto a carrier wave. A basic modulator circuit has three connections: an input for the unmodulated carrier, an input for the signal, and the modulated output. The types of modulation have been discussed above. Some important additional points to note about each type are highlighted below.

- **Amplitude Modulation (AM):** AM signals contain a central carrier frequency and two sidebands. The USB is the sum of the carrier and audio frequencies, while the LSB is the difference. There are some variations to this principal structure:

    - **Double Sideband Suppressed Carrier (DSB-SC)**: Here, the carrier is suppressed for efficiency, and only the sidebands are transmitted.

    - **Single Sideband (SSB)**: Here, only one sideband (USB or LSB) is transmitted. This further optimizes power usage and spectrum efficiency. SSB can be generated with filter or phase techniques. These will be discussed in later sections.

- **Frequency Modulation (FM) and Phase Modulation (PM):** Although these two modes are technically distinct, they often sound indistinguishable on air and can be demodulated through the use of similar circuits. A common method to produce FM or PM is the reactance modulator. When connected to an oscillator's frequency control circuit, it induces frequency shifts (FM). In contrast, when connected to an RF amplifier stage after the oscillator, it causes phase shifts (PM).

- **Quadrature (I-Q) Modulation:** Also known as I/Q modulation, this technique can be used to transmit digital data and has the flexibility to synthesize various modulation types. The "I" (in-phase) and "Q" (quadrature) refer to signals that have the same frequency but a 90-degree phase offset. The versatility of quadrature modulation makes it ideal for software-defined radios (SDRs), which use digital signal processing. I-Q modulation involves a few key steps:
  - **Carrier Manipulation**: The carrier signal is split into two, and one branch undergoes a 90-degree phase shift.
  - **Mixing**: The original and phase-shifted carriers are individually mixed with I and Q data-carrying signals.
  - **Combination**: The resultant mixed signals are combined. The final modulated output contains two signals with carriers 90 degrees out of phase.

Digital Signal Processing (DSP) in Software-Defined Radios (SDRs)

DSP is a powerful technique used to manipulate signals within radio systems. It is used in radio systems for noise reduction, demodulation of various modes, and other signal enhancement functions.

Stages of Digital Signal Processing

DSP involves a series of conversions and computations:

- **Analog to Digital Conversion (ADC)**: The received RF signal, either directly or after being downconverted to an intermediate frequency (IF), is sampled and converted into digital data by an ADC.
- **Digital Processing**: A specialized microprocessor, the DSP, performs complex mathematical operations on the digital signal. This is the heart of the system, where functions like noise reduction and demodulation can occur.
- **Digital-to-Analog Conversion (DAC)**: The processed digital data is converted back to analog form by a DAC. This analog output can be audio for the operator, or it may be further processed for digital modes.

Importance of Digital Signal Processing

DSP enhances performance and makes radio systems highly adaptable. DSP receivers provide significant advantages compared to traditional analog designs, such as highlighted below:

- **Preprogrammed Filters**: DSP receivers often come with a selection of built-in filter profiles tailored to common modes (for example, SSB, CW, FM).
- **Adjustable Filters**: Operators can often fine-tune the bandwidth and shape (steepness of the filter edges) of filters to best isolate the desired signal and minimize interference.
- **Customizable Filters**: Advanced DSP receivers may even allow users to create entirely new filter profiles. This capacity affords hams the utmost control over how the receiver processes incoming signals.

Nevertheless, too much processing can cause signal distortion.

Switch-Mode Power Supplies

Switch-mode power supplies use transistors to convert electrical power efficiently. Unlike linear power supplies, which rely on bulky transformers to regulate voltage, switch- mode power supplies employ transistors to rapidly switch current pulses at high frequencies (usually between 20 kHz and 500 kHz) through a transformer and/or inductor. A capacitor then filters the voltage to produce a smooth DC output.

Advantages of Switch-Mode Power Supplies

- **Size and Weight Reduction:** Switch-mode power supplies operate at high frequencies. Therefore, they can use smaller transformers and inductors than linear supplies. This causes appreciable reductions in size and weight.
- **Efficiency:** Switch-mode power supplies are generally more efficient than linear supplies. Since they rely on switches rather than linear voltage regulation, they waste less energy as heat.

Comparison to Linear Power Supplies

Linear power supplies, in contrast, depend on bulky transformers with iron cores that operate at the standard line frequency (50 or 60 Hz). These transformers contribute to the larger size and weight of linear power supplies.

Connectors

Connectors are essential components that establish secure physical and electrical connections within an electronic system. They are generally categorized as:

- **Plugs**: Plugs are designed for termination on cables.
- **Jacks**: Jacks are fixed connectors attached to equipment or devices.
- **Adapters**: Adapters facilitate compatibility between different connector types.
- **Splitters**: Splitters enable the division of a signal into multiple paths.
- **Keyed Connectors**: Keyed mechanisms prevent incorrect or reversed connections.

Audio and Control Connectors

Numerous connector types are used in consumer electronics and amateur radio for audio and control signals. Common examples include:

- **RCA Phono Connectors**: These versatile connectors are widely used to carry low-frequency audio and video signals. Phono connectors can be used to transmit audio signals but are not usually used as DC power connectors for radio equipment.
- **DIN/Mini-DIN Connectors**: DIN and mini-DIN connectors have a keyed interface with multiple pins. This makes them well-suited for audio signals and various control functions within radio systems.
- **Tip-Ring-Sleeve (TRS) Connectors**: Certain audio connectors, like the common one-quarter phone jack, are arranged in a tip-ring-sleeve (TRS) configuration. The tip at the end of the plug carries the main signal component (for example, the positive signal in mono audio). The ring often carries an additional signal channel (for example, the negative signal in mono audio or one channel of a stereo signal). The sleeve functions as the circuit's electrical ground (base) or reference.

Anderson Powerpole Power Connectors

Anderson Powerpole Connectors have become widely adopted within the Amateur Radio community, especially in ARES applications. Their popularity stems from their robustness, ease of use, and ability to prevent accidental polarity reversal.

RF Connectors

Specialized connectors are designed to handle RF signals. Common types include:

- **UHF Connectors (SO-239 & PL-259)**: These are used in the HF and lower VHF frequency ranges (up to around 150 MHz).

- **"N" Connectors**: These are threaded RF connectors known for their durability and moisture resistance. N connectors can operate effectively at frequencies up to approximately 10 GHz. They are commonly used in professional and higher-power radio applications where reliability in harsh environments is important.

- **SMA (SubMiniature version A) Connectors**: These compact, threaded connectors are designed for use with miniature coaxial cables. They have a small size and threaded design for secure connections and are suitable for frequencies up to several gigahertz (GHz). SMA connectors are often found on antennas, RF test equipment, and modern radio transceivers.

- **BNC (Bayonet Neill–Concelman) Connectors**: BNC connectors are popular for their quick-connect/disconnect mechanism. They have a bayonet-style lock system for fast and secure connections. BNC connectors are usable up to around 2 GHz for optimal low-SWR performance. An SWR below 1.5:1 is generally considered good for most applications.

Data Connectors

- D-Type Connectors: D-Type connectors (like DE-9 and DB-25) are used for serial (RS-232) and parallel data connections.

- USB Ports: USB ports have become the modern standard to connect computers to peripheral devices (including many contemporary radio transceivers).

Circuit Symbols

Field Effect Transistor

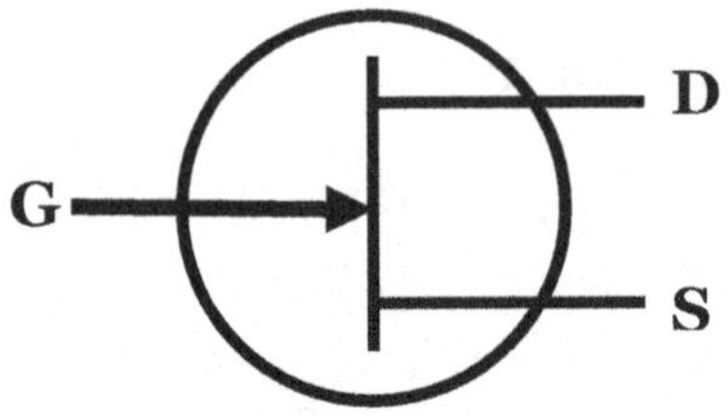

NPN Junction Transistor

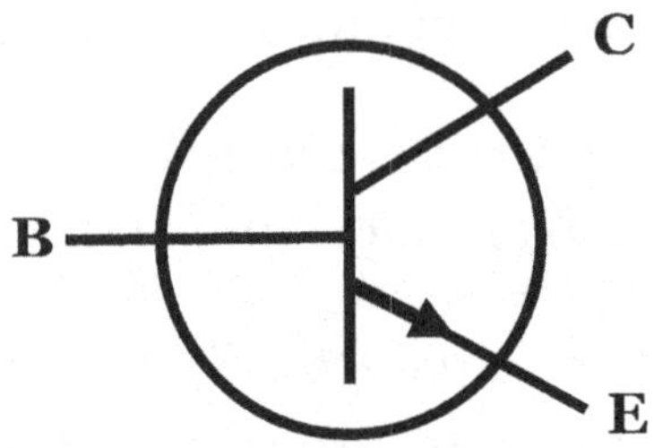

Solid-Core Transformer

Zener Diode

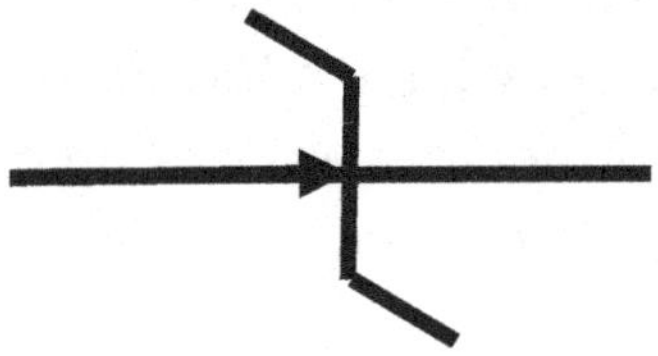

Tapped Inductor

Chapter 8: Signals and Emissions

The world of radio communication teems with unseen signals that carry voices, data, and images across vast distances. This chapter unpacks the fundamental principles of radio waves, their behavior, and the specialized equipment designed to harness their power. From basic transmitters and receivers to sophisticated antennas and signal-processing tools, we will explore the technologies that transform silent frequencies into meaningful communication.

Basic Radio Modes

Modulation: The Foundation of Information Transmission

Modulation encodes information onto an RF carrier wave for transmission. This enables information such as voice, music, or data to be carried over long distances via radio signals. Imagine a carrier wave as a highway and the information you want to transmit as a car. Modulation places the car on the highway. There are different ways to modulate a carrier wave, and each depends on what kind of information you want to transmit. Three common modulation modes used in radio transmission are discussed below.

- **Amplitude Modulation (AM)**: In AM mode, the power (strength or amplitude) of the RF carrier wave is varied in proportion to the audio signal being transmitted. Imagine a radio wave as a wave on the ocean. In AM mode, the information signal (like your voice) controls the height of the wave (amplitude). When your voice gets louder, the wave gets taller. When your voice gets quieter, the wave gets shorter. The varied peak values of the AM signal's waveform create an "envelope" that contains the encoded audio information. The original audio information can be recovered from this modulated waveform through a process called detection, which extracts the variations in the signal's amplitude envelope. AM is still used for some long-distance radio communication applications but is highly susceptible to noise interference.

- **Frequency Modulation (FM)**: In FM mode, the frequency of the RF carrier is shifted slightly higher or lower in accordance with the audio signal. Imagine the radio wave as a car on a racetrack. In FM, the information signal (like your voice) controls the speed of the car (frequency). When your voice gets louder, the car speeds up slightly. When your voice gets quieter, the car slows down slightly. The frequency of the radio wave changes slightly based on the information. FM radio stations use this method, which is less susceptible to noise interference.

- **Phase Modulation (PM)**: The PM mode is related to frequency modulation. It involves a change in the phase angle of the RF carrier to represent the information being conveyed. Imagine an analog clock. Its hands move smoothly with precise intervals. In phase modulation, the information signal slightly advances or delays the "ticks" of these hands. When the movement of one of the two hands is delayed, you can increase the angle between the hands. The carrier wave's regular oscillations are subtly shifted forward or backward in time based on the information conveyed. PM is often used in combination with frequency modulation or as part of digital communication systems.

Bandwidth

In communication systems, bandwidth refers to the range of frequencies occupied by a signal. It is usually measured in Hertz (Hz) and represents the difference between the highest and lowest frequencies in the signal. The table below is a guide to the common signal bandwidths accessible by ham operators.

Signal Type	**Bandwidth**
AM Voice.	6 kHz.
Amateur Television.	6 MHz.
CW (Morse).	100 Hz to 300 Hz.
Digital (with SSB).	50 Hz to 3 kHz.
FM Voice.	5 kHz to 16 kHz.
SSB Voice.	2 to 3 kHz.

Radio Path Losses

Radio signal transmission is not without its challenges. Between the transmitter and receiver, various factors can degrade signal strength. These factors are collectively known as radio path losses.

Factors that Contribute to Path Losses

These factors contribute to path loss, each with its unique impact, are discussed below:

- **Free Space Loss**: This is the loss experienced by any radio wave due to the natural dissipation of its energy as it travels through space. The farther the signal travels, the weaker it becomes. The signal strength follows an inverse-square law relationship with distance.

- **Diffraction**: Radio waves can bend slightly around obstacles like buildings, hills, or even the curvature of the earth. Although diffraction can help extend signal coverage in some cases, it can also weaken a signal, especially when obstacles are large relative to the wavelength.

- **Multipath**: In a non-ideal environment, radio waves can reflect off various surfaces like buildings, terrain features, or even airplanes. These reflections create multiple paths for the signal to reach the receiver. However, these paths have slightly different lengths that can cause the signals to arrive at the receiver slightly out of sync. This can produce two phenomena:

    - **Constructive interference**: When the reflected and direct waves are in phase (peaks align), the signal strength at the receiver can be amplified.

    - **Destructive interference**: When the reflected and direct waves are out of phase (peaks and troughs do not align), the signal strength can be significantly weakened or even completely canceled out in extreme cases. The overall effect of multipath on signal strength depends on the specific environment and the relative phases of the signals.

- **Absorption**: Certain materials, such as water vapor in the atmosphere, foliage, or construction materials, can absorb radio wave energy. This absorption directly contributes to signal loss, especially at higher frequencies where water vapor has a more significant impact.

- **Terrain**: The nature of the terrain between the transmitter and receiver can significantly influence signal propagation. Mountains, valleys, and large bodies of water can act as barriers that block or attenuate the signal.

- **Atmosphere**: Although generally transparent to lower-frequency radio waves, the earth's atmosphere can affect higher-frequency signals through phenomena like refraction and scatter. Refraction bends the radio waves as they pass through layers of air with varied densities. Scatter occurs when radio waves interact with particles in the atmosphere that cause them to deviate from their original path. These atmospheric effects can cause signals to fade, distort, or even be lost completely.

System Gain

System gain represents the net increase in signal strength across a radio communication link, considering both the transmitter's output power and the receiver's sensitivity, along with any gains or losses from antennas, amplifiers, and cables. The minimum required system gain must be determined to ensure that the signal reaches the receiver with sufficient strength for reliable detection.

Imagine that you need to shout across a vast distance. Without amplification, your voice will be weak and easily lost. Similarly, radio waves naturally weaken as they travel due to factors like dispersion in space (free space loss) and obstacles. System gain acts as the amplification mechanism in a radio system that ensures the signal reaches the receiver with sufficient strength for proper demodulation and clear communication.

Some of the factors that contribute to system gain include:

- **Transmitter Power**: The raw power output of the transmitter, usually measured in watts (W) or in decibels relative to one milliwatt (dBm). Higher transmitter power translates to a stronger initial signal.

- **Transmit Antenna Gain**: Antennas can focus radio waves in specific directions and amplify the signal strength in those directions. This gain is measured in decibels relative to an isotropic antenna (dBi), which radiates equally in all directions.

- **Receive Antenna Gain**: To amplify the signal, the receiver antenna can focus on radio waves. This gain also contributes to the overall system gain.

To calculate system gain, add the individual gains (in dB) and subtract the losses (also in dB) encountered along the path:

- System Gain (dB) = Transmitter Power (dBm) + Transmit Antenna Gain (dBi) + Receive Antenna Gain (dBi) – System Losses (dB).

Link Margin

Link margin, often expressed in dB, represents a safety buffer built into the system design. It is also known as "fade margin" or "excess path loss." It accounts for potential signal fade due to environmental factors or variations in propagation conditions. Adequate link margin is vital to maintain reliable communication, especially with digital signals, which can be more sensitive to signal strength fluctuations.

Link Budget

A link budget provides a comprehensive analysis of all power gains and losses within a radio system. A link budget meticulously accounts for all the factors that impact signal strength at each stage of the communication link. This systematic approach helps ensure the received signal power meets the minimum threshold for successful demodulation. Factors usually considered in the link budget include system gains, system losses, and link margin.

Transmitters

Since the early days of spark-gap transmitters and AM, radio technology has advanced dramatically. In this section, we will examine the methods behind CW, SSB, and FM. Whether your equipment uses traditional hardware or software-defined radio, this knowledge will help you craft a clean, effective signal.

FM Transmitters

Harmonics in FM Transmission

FM transmission produces a low-frequency signal, which is then frequency-multiplied to the desired transmission band. For example, to transmit on the 70 cm band at 442.00 MHz, the initial FM signal might be at 11.05 MHz, which is then multiplied by a factor of 40.

Frequency Deviation and Harmonics

Think of FM modulation as a rope that wiggles. The speed of the up-down wiggle is like the audio frequency. Faster up-down wiggles mean higher-frequency sounds. How far you wiggle the rope side-to-side represents the amount of frequency deviation.

The frequency deviation of the FM signal is also multiplied along with the carrier frequency. Now, imagine the rope is attached to a machine: the harmonic multiplier. The machine stretches the rope and makes it longer (this correlates to how someone would multiply the frequency to reach 442.00 MHz). When you stretch the rope, the wiggles get bigger too. This means the deviation increases along with the frequency.

The frequency deviation (Δf) of a reactance-modulated oscillator in an FM transmitter can be calculated with the formula $\Delta f = (\Delta F\_max) \div (f\_out \div f\_in)$. In the formula, Δf = Frequency deviation (Hz), ΔF_max = Maximum deviation frequency (Hz), f_out = Output frequency of the FM transmitter (Hz), and f_in = Input frequency of the reactance modulated oscillator (Hz).

In the given example, $\Delta f\_out = 4000$ Hz (desired deviation on the 442.00 MHz signal), $f\_out = 442$ MHz $= 442 \times 10^6$ Hz, and $f\_in = 11.05$ MHz $= 11.05 \times 10^6$ Hz. When we substitute the values into the formula, we get:

- $\Delta f\_in = 4000$ Hz $\div (442 \times 10^6$ Hz $\div 11.05 \times 10^6$ Hz$)$.
- $\Delta f\_in = 4000$ Hz $\div 40$.
- $\Delta f\_in = 100$ Hz.

Therefore, the frequency deviation on the original 11.05 MHz signal must not exceed 100 Hz to ensure that the final 442.00 MHz signal stays within its allowed bandwidth. To keep the FM signal within its allowed bandwidth of 442.00 MHz, you need to start with smaller side-to-side wiggles (less deviation) at the start. That is why the original 11.05 MHz signal can only have a 100 Hz deviation instead of the full 4 kHz.

Bandwidth Considerations and Carson's Rule

The FCC mandates that FM transmissions adhere to good amateur practices when it comes to bandwidth. FM signals do not have sharp edges like AM signals. To figure out how much space an FM transmission takes up, Carson's Rule gives us a good estimate: *Bandwidth (BW) = 2 × (Maximum Deviation + Highest Audio Modulation Frequency).*

Maximum (peak) deviation refers to how far the carrier frequency swings back and forth due to the modulation. A wider swing means a wider signal. The highest audio modulation frequency is the highest pitch in the audio signal transmitted. Higher pitches change the carrier frequency more quickly and spread the signal wider. We multiply by 2 because FM signals spread out on both sides of the carrier frequency, so we need to double our calculation to get the total bandwidth.

For example, if your FM radio has a maximum deviation of 3 kHz and you transmit voice with frequencies up to 2.5 kHz, the estimated bandwidth would be BW = 2 × (3 kHz + 2.5 kHz) = 11 kHz.

SSB Transmitters

The quest for efficient and clear voice transmission led to the development of SSB transmitters. The balanced modulator is the vital first stage in an SSB transmitter. It suppresses the carrier and generates two sidebands. To isolate a single sideband for transmission, either a filter or phase method can be employed.

Balanced Modulators

Balanced modulators are fundamental to SSB signals because they efficiently suppress the carrier wave. This improves power efficiency and reduces bandwidth. A balanced modulator is a circuit that combines two input signals (the carrier signal and the audio signal) to generate a double-sideband suppressed carrier (DSB SC) signal. The steps a balanced modulator undergoes are discussed below:

- **Carrier Signal Splitting:** The carrier signal from the oscillator is fed into the balanced modulator circuit. Internally, this carrier is divided into two equal paths.

- **Phase Inversion in One Path:** One of the carrier signal paths experiences a 180-degree phase shift (inversion). This means the peak of the wave becomes the trough, and vice versa.

- **Balanced Mixing with the Modulating Signal:** The original and phase-shifted carrier paths are each combined with the modulating signal (usually audio from your microphone) through the use of transistors or diodes. The way these components are arranged allows the product (mixing) of the carrier and modulating signal to occur differentially.

- **Balanced Cancellation of the Carrier:** Due to the carefully matched components and the phase inversion in one path, the carrier signal itself cancels out when the two mixed outputs are summed.

- **Double-Sideband Output:** The resultant output from the balanced modulator is a DSB SC signal. It contains the USB and LSB, which each represent the sum or difference of the original carrier frequency and the modulating frequency. If fc represents the carrier frequency and fa represents the modulating (audio) frequency, the double-sideband output contains the upper sideband (fc + fa) and the lower sideband (fc – fa).

Two Methods to Generate SSB

To transmit SSB, the desired sideband (USB or LSB) still needs to be selected from the double sideband output. This selection can be achieved through the use of one of the methods discussed below.

- **The Phasing Method**: The phasing method produces exceptional sideband suppression, but hams find it more complex to achieve in practice. It involves precise manipulation of the phase of the carrier signal. Two balanced modulators are used, each of which receives a version of the carrier with one signal shifted by 90 degrees relative to the other. The outputs of these balanced modulators are then carefully combined to cancel out the unwanted sideband. Historically, the method was pioneered by Central Electronics in the 1950s. This method is very sensitive to drift in vacuum tube-based circuits and needs to be frequently retuned, which reduces its popularity.

- **The Filter Method**: This simple approach relies on selective filters to isolate the desired sideband. After the balanced modulator generates a DSB SC signal, a high-quality filter with a sharp cutoff frequency is used. This filter acts like a gate that allows either the USB or LSB to pass through while frequencies are attenuated on the opposite side.

Distortion in SSB Circuits

Distortion in SSB circuits is any unwanted alteration of the transmitted signal that deviates from a faithful representation of the original audio input. Non-linearities in amplifiers, mixers, or filters can introduce harmonics and intermodulation products that degrade the signal and create spurious emissions outside the intended sideband. Furthermore, imbalances in frequency or phase response through the SSB circuit may distort the audio content of the signal and reduce clarity. To maintain a clean SSB signal, careful attention must be paid to the linearity of components (especially in amplifiers) and proper filter design.

Ensure Signal Fidelity

Transmission of a clean, intelligible signal that remains within allocated bandwidth limits is paramount for responsible amateur radio operation. This minimizes interference and ensures maximum communication effectiveness. Some of the key factors that impact signal fidelity are discussed below.

Overmodulation

Overmodulation is a critical issue in ham radio transmission, especially for AM and SSB modes. It occurs when the modulating signal (usually audio from the microphone) becomes too strong relative to the carrier wave and causes distortion and unwanted signal products. Here is a breakdown of the important concepts related to overmodulation:

- **Modulation Envelope:** Imagine the outline of your transmitted signal as it fluctuates with the audio content. This outline is called the modulation envelope. In proper modulation, the envelope smoothly follows the ups and downs of your voice.

- **Flattopping:** Flattopping happens when your transmitter reaches the limits of its power output. Even if your voice gets louder (the modulating signal increases), the amplified signal is "clipped," and its peaks appear flattened. This signifies the difficulty the amplifier has with the strong audio signal.

- **Carrier Cutoff**: This is an extreme form of overmodulation where the signal is not merely clipped, but completely cut off for brief intervals, especially between voice peaks. This cuts off the carrier frequency itself. Like flattopping, carrier cutoff is a sign the transmitter is being pushed too hard. Carrier cutoff produces even more severe distortion than regular flattopping.

- **Distortion Products:** This abrupt clipping is a major source of distortion. The distortion creates unwanted signals (harmonics and other products) that can spread far outside your intended transmission bandwidth and contribute to splatter and buckshot.

- **Splatter and Buckshot:** These terms describe the broad splatter of unwanted emissions caused by overmodulation. Imagine that you throw a handful of pebbles (your original signal). Some land where intended (your sidebands), but others scatter widely (distortion products) and cause interference across the band.

- **Automatic Level Control (ALC):** Many modern transceivers incorporate ALC circuits. ALC acts like a safety valve that automatically reduces the gain of the audio signal before it clips the amplifier and causes overmodulation. Proper ALC settings help maintain a clean signal.

- **Microphone Gain:** The microphone gain control adjusts the strength of the audio signal that enters the transmitter. A gain set too high is a common cause of overmodulation. A good practice is to adjust the gain so that ALC activates only on occasional voice peaks.

Two-Tone Testing for Linearity

Two-tone testing can be used to assess how well your transmitter (and any external amplifiers) handle complex signals and reveal potential distortion. It evaluates the linearity of the transmitter and any external amplifiers used in AM or SSB transmissions. The processes involved in two-tone testing are discussed below.

- **Signal Generation**: Two audio tones with harmonically unrelated frequencies are introduced into the transmitter's audio input simultaneously.
- **Monitor the Output**: As these tones are transmitted, the signal that results is monitored on a monitoring oscilloscope.
- **Interpret the Results**: In an ideal scenario, a linear system would reproduce these tones precisely on the oscilloscope. However, any non-linearities within the transmitter or amplifier will introduce distortion products. These distortions manifest as additional frequencies that appear on the oscilloscope display.

Amplifier Efficiency and Classes

Efficiency

Mathematically, the efficiency of a machine is given as (Energy Output ÷ Energy Input) × 100. Amplifier efficiency refers to how much RF output power is achieved relative to the DC input power consumed—that is, the percentage of DC input power successfully converted into usable RF output power. The higher the ratio of output to input, the more efficient the amplifier. Efficiency is especially important for battery-powered or portable operations.

Amplifiers are categorized into classes (A, B, AB, and C) based on how much of the input signal waveform they amplify. This "biasing" significantly impacts power efficiency. Linear amplifiers are designed to process the complete input waveform (both positive and negative cycles). This ensures the output signal closely mirrors the input and minimizes distortion.

Amplifier Classes and Waveform-Power Trade-Offs

There is an inherent trade-off between efficiency and linearity in amplifier design. The type that is needed depends on the mode of operation. SSB and AM prioritize linearity and generally require Class A or AB amplifiers. FM and CW can tolerate greater distortion but require the higher efficiency that Class C amplifiers provide.

- **Class A** amplifies the entire input waveform continuously. This produces very minimal distortion (that is, great linearity). However, power efficiency is the lowest, as the amplifier draws DC power constantly.

- **Class B** uses two devices (often transistors) in a "push-pull" configuration. Each device amplifies only half of the input waveform and switches on and off in turn. This reduces DC power consumption for roughly 50% of the time.

- **Class AB** is a more efficient but less linear option than Class A. It compromises between A and B to amplify slightly more than half of the waveform with moderate distortion.

- **Class C** prioritizes efficiency above all else. It amplifies a very small portion of the input waveform. Although highly power efficient, Class C amplifiers introduce significant distortions into the output signal.

Transmit (TX) Delay Adjustments

When a transceiver is used to drive a linear amplifier, a TX (transmit) Delay is often necessary. This is because of the inherent delay between the moment the PTT (push-to-talk) circuit is activated, and the instant both the transceiver and the amplifier are fully ready to transmit. The exact delay depends on the switching mechanisms used.

If a transceiver starts to transmit RF energy before the amplifier is fully switched into transmit mode, it can cause "hot switching." This can cause arcing and component failures. The TX Delay introduces a deliberate pause after the PTT circuit is activated. This ensures that the amplifier has fully transitioned to transmit mode before a transceiver starts to send RF power.

Tube Amplifier Neutralization

Vacuum tubes harbor capacitance between their internal elements (like the grid and plate). This capacitance can create a pathway for positive feedback within the amplifier. If strong enough, this feedback can cause self-oscillation at very high frequencies—a dangerous and uncontrolled condition.

Neutralization combats self-oscillation through the deliberate introduction of a small, out-of-phase signal that precisely counteracts the problematic positive feedback. This suppresses those unwanted high-frequency oscillations and restores stability, which effectively neutralizes the amplifier.

Tube Amplifier Tuning

To tune a tube amplifier, you must adjust the output circuit to resonate at your operating frequency. This resonance ensures that energy from your transmitter is efficiently transferred to the antenna and maximizes radiated signal power. A mismatched, non-resonant amplifier wastes power, weakens your signal and generates excess heat in the tube. The steps to manually tune an amplifier are discussed below.

- **Band Selection**: Ensure the BAND switch is set to the frequency range you want to operate within.
- **Initial Resonance**: Start with a low drive power level. Adjust the plate capacitor until you observe a distinct "dip" in plate current. This indicates that the output circuit is resonant at your desired frequency.
- **Maximize Output**: Carefully adjust the loading capacitor to achieve maximum output power. Always stay within the tube's maximum plate dissipation rating to avoid damage.
- **Refinement**: If necessary, slightly increase the drive power and repeat the plate capacitor adjustment for another dip in plate current.
- **Final Adjustment**: Once again, tweak the loading capacitor for maximum output and ensure safe plate current levels.

Carefully monitor the grid current and ensure it never exceeds the tube's maximum rating. Excessive grid current is a sign of overdrive, which can rapidly damage the tube.

Receivers

Superheterodyne Receivers

Most modern receivers use the superheterodyne (superhet) design, where incoming signals are mixed with an LO to produce a fixed IF. This simplifies high-performance filtering and amplification.

The Simplest Superhet

At its most basic level, a superheterodyne receiver can be built with just three key elements:

- **Local Oscillator (LO)**: This generates a tunable internal signal. The LO's frequency is carefully controlled to mix with the incoming radio signal.

- **Mixer**: This circuit combines the incoming RF signal with the signal from the local oscillator. This process is what gives the superheterodyne design its name and creates a new signal at a fixed intermediate frequency (IF).

- **Detector**: When operated directly at the IF, the detector extracts the desired information from the signal. In SSB and CW devices, this involves the use of a **product detector** to combine the IF signal with a locally generated signal to produce an output in the audio frequency range. In AM devices, a product detector (for superior performance) or an **envelope detector** (simpler design) can be used. Unlike product detectors, envelope detectors extract the audio signal through the use of the amplitude variations of the AM carrier wave. In FM receivers, **discriminators** or **quadrature detectors** convert FM frequency shifts into an audio signal that represents the original modulation.

Single-Conversion Superhet

Single conversion superheterodyne receivers provide more sophisticated systems for better signal processing. The stages for the single conversion superhet are as follows:

- **RF Amplifier**: This is the initial boost of the weak incoming radio signal.

- **Mixer**: This combines the amplified RF signal with the LO signal to produce a mix of frequencies, including a fixed IF. Both the RF amplifier and the mixer are also collectively called the front end.

- **IF Filter**: This narrows the receiver's focus to a specific band of frequencies centered around the desired IF.

- **IF Amplifier**: This significantly increases the signal strength within the filtered IF band. This is where most of the receiver's gain occurs.

- **Demodulator**: This extracts the intended information from the IF signal. The type of demodulator depends on the mode (such as AM, SSB, and FM).

- **Audio Amplifier**: The final amplification stage drives the headphones or speaker.

Double-Conversion Superhet

The double conversion introduces an extra mixing stage and an additional IF section. This produces greater selectivity (narrower filtering) and improved image rejection by filtering at multiple frequencies.

- **RF Amplifier**: This is an initial boost of the weak incoming signal, just like in a single conversion design.
- **Mixer 1**: This mixes the RF signal with the first local oscillator (LO1) to produce a primary intermediate frequency (IF1).
- **IF1 Filter**: This isolates a band of frequencies centered on IF1.
- **Mixer 2**: This combines the filtered IF1 signal with a second local oscillator (LO2) to generate a second, lower intermediate frequency (IF2).
- **IF2 Filter**: This provides a second (often narrower) filtering stage centered on IF2.
- **IF2 Amplifier**: This provides a significant gain increase for the IF2 signal.
- **Demodulator**: This extracts the audio or data information encoded in the IF2 signal.
- **Audio Amplifier**: This is the final amplification stage for speakers or headphones.

Challenges of Superhet Receivers

Though they are the cornerstone of modern radio design, superhets face potential challenges that can hinder signal clarity. One of the most important drawbacks of superhet receivers—image response—is discussed below.

Image Response: In a superheterodyne receiver, the mixing process creates both sum and difference frequencies. This means an undesired signal, located at a specific offset from your intended frequency, can also create the desired IF frequency. This unwanted signal is called an image response. Strong front-end filters, often placed before the initial mixer, help to minimize the risk of image responses that interfere with reception. To illustrate the concept of image response, consider the following example:

- **IF**: This is a fixed frequency within the receiver chosen for optimal filtering and amplification. Assume this frequency is 10.7 MHz.
- **LO**: This is the tunable oscillator frequency within the receiver. Assume this frequency is 28 MHz.
- **Desired Signal**: This is the radio signal you want to receive. Assume this frequency is 38.7 MHz.

- **Image Frequency**: The key here is the image frequency. This is the unwanted signal that can create the same IF as your desired signal. We can calculate it with the formula: Image Frequency = LO frequency + IF.

- IF = 28.000 MHz (LO) + 10.7 MHz (IF) = 38.700 MHz.

Notice how the IF (14.255 MHz) is the same as our desired signal frequency! This creates the potential for interference. To minimize the impact of image responses, receiver designers employ filters, usually located before the first mixer stage. These filters aim to attenuate strong signals at the image frequency so they don't mix with the LO, which creates interference.

How to Balance Gain, S-Meters, and Linearity

Receivers must amplify weak incoming signals to make them audible or usable by computers. However, too much gain has its drawbacks. Gain can be controlled manually or automatically.

- **Manual RF Gain Control**: This provides direct control over the initial amplification stages. Maximum RF gain is useful when you want to find weak signals, but for normal reception, a reduced setting can minimize background noise.

- **Automatic Gain Control (AGC)**: Radio signals can vary greatly in strength due to fading, distance, and other factors. AGC circuits within a receiver automatically adjust their sensitivity to maintain a relatively constant audio output level at the speaker or headphones. AGC tames strong signals and prevents them from overload and audio distortion. It also boosts weak signals and makes them easier to hear. Many receivers allow you to fine-tune the AGC's response time. A slow AGC setting is ideal for voice modes (like SSB), where signal fluctuations occur naturally. For CW and digital modes, a fast AGC helps the receiver quickly adapt to the rapid on/off nature of the signal.

Receiver Attenuators

Receiver attenuators perform a specific function. They enhance the receiver's ability to manage strong overload signals without distortion. Attenuators reduce the strength of incoming signals before they reach the most sensitive stages of the receiver.

Manage Noise

Modern receivers incorporate a selection of tools to enhance signal clarity. For ideal reception, the receiver's bandwidth should closely match the bandwidth of the transmitted signal. This maximizes the signal-to-noise ratio and makes it easier to discern the desired signal from background noise. Core IF filters are the first stage to help radio operators achieve this. However, beyond the core IF filters, some receivers incorporate additional tools and techniques to address specific interference.

- **Notch Filters**: As discussed earlier, these highly adjustable filters target and suppress narrowband interference, such as a persistent carrier wave. They can operate at audio or IF frequencies.

- **IF Shift (Passband Tuning)**: This feature subtly shifts the receiver's passband slightly higher or lower in frequency, while the displayed carrier frequency remains unchanged. This slight offset can sometimes be enough to improve signal isolation.

- **CW-R (Reverse Sideband)**: To receive CW, some receivers allow you to select either normal or reverse CW operation. This helps avoid interference from adjacent CW signals. When you switch between USB and LSB, this effectively repositions your receiver's focus. The interference, which was previously close to your desired signal within the receiver's passband, might now be shifted to the opposite sideband. This allows the receiver's filters to more effectively suppress the interference since their response usually slopes away from the center frequency.

- **Noise Blankers and Noise Reducers**: These circuits specifically target impulsive noise (like ignition noise), which temporarily mutes the receiver during the noise pulses. This feature should not be overused, as it can distort strong signals.

Basics of Digital Modes

Digital modes represent a unique way to transmit information over radio waves. They encode text, data, or even digitized voice into patterns of tones or other signal changes within a traditional voice channel (like SSB). These modes, like FT8, RTTY, and PSK31, are advantageous for weak signal communication and automated information transfer. Some digital modes, such as D-STAR and System Fusion, specifically transmit compressed voice.

Sound Cards: Convert Data to Transmitted Signals

Digital modes bridge two worlds: the digital realm of your computer (where information exists as ones and zeros) and the analog world of radio transmissions (represented by audio tones). Specialized software, in conjunction with your computer's sound card, transforms digital information (ones and zeros) into audible tones or other modulation techniques suitable for your radio.

When you play a CD or record audio with a microphone on your computer, the sound card handles the conversion between the analog audio signals and digital data. The same principle applies to digital radio modes. To direct the audio flow to and from your radio, specialized audio driver software is essential. It acts as a controller that routes signals between your digital mode programs and the sound card.

The digital mode setup provides flexibility where DAC and ADC conversions occur. Here are the common locations:

- **Within the Computer**: Desktop and laptop computers often include built-in sound cards as separate cards or integrated into the main motherboard.
- **External Interface**: Devices like the RigExpert or Signalink provide dedicated sound cards in an external box. These are popular for their straightforward connections and physical adjustment controls.
- **Inside the Radio**: Some newer transceivers feature integrated sound cards. This simplifies the setup, as only a USB connection is needed between the radio and the computer.

Regardless of where the sound card is located, an audio driver software is still needed on the computer to manage the audio flow and interface with chosen digital mode programs.

Find Digital Activity

Digital modes are usually found within the data/CW portions of specific HF bands. The following table highlights some common areas for digital activity.

HF Band	Frequency Range (MHz)	Special Notes
160 meters.	1.800–1.810.	
80 meters.	3.570–3.600.	
60 meters.	FCC-specified channels.	Consult FCC regulations for details.
40 meters.	7.070–7.125.	7.040 MHz: RTTY DX call frequency.
30 meters.	10.130–10.150.	
20 meters.	14.070–14.0995, and 14.1005–14.112.	14.070 MHz: PSK31 call frequency.
17 meters.	18.100–18.110.	
15 meters.	21.070–21.110.	
12 meters.	24.920–24.930.	
10 meters.	28.070–28.189.	

Digital Voice Modes

Common choices on the VHF and UHF bands include proprietary digital voice systems. Among the most popular are DMR (developed by Motorola), D-Star (Icom), and System Fusion (Yaesu). These standards provide enhanced audio quality and advanced features compared to traditional analog voice modes. Digital voice modes also exist for HF operation. AOR's Digital Voice System (commercial) and FreeDV (open-source) are common examples.

Character-Based Modes

Character modes, often called "keyboard-to-keyboard," facilitate direct text-based communication between radio operators. For example, one of the oldest digital modes, RTTY, initially relied on mechanical teleprinters. It has since evolved to use computers and sound cards and remains active on the HF bands. Similarly, PSK31 is a popular digital mode known for its efficient use of bandwidth and strong performance under weak signal conditions.

Definitions of Important Terms

- **Bit**: The most basic unit of digital data, represented by either a 0 or a 1.
- **Bit Rate**: The speed at which digital information is transmitted, expressed as the number of bits sent per second.
- **Symbol (Baud)**: One distinct change to a transmitted signal that carries data. The symbol rate is measured in baud.
- **Duty Cycle**: The percentage of time a transmitter actively transmits compared to its total operational cycle (both on and off time).
- **Protocol**: A set of rules that define how data is exchanged between devices in a digital mode.
- **Mode**: The specific combination of a protocol (like RTTY) with a method of signal modulation (such as FSK).

Baud Versus Bit Rate

Imagine that you use your hand to signal information. Each distinct position (open hand, one finger extended, two fingers, and so on) represents a symbol. In this analogy, if you can change your hand position once per second, your symbol rate (baud) is one.

Now, assume each hand position represents a unique five-digit binary number. Since there are 2^5 (32) possible combinations (including all zeroes), a single symbol change can convey the equivalent of five bits of information. Therefore, with a one-symbol-per-second limit, your bit rate would be five bits per second.

In radio transmissions, symbols are represented by factors like signal phase shifts, distinct tones, or precise voltage levels. The FCC regulates the maximum symbol rate (baud) allowed for digital modes on specific bands.

Baud and Bits/Second

Digital signals can be designed where a single change or symbol (measured in baud) represents multiple bits of information. This occurs when each symbol can exist in more than just two distinct states (baud). Consider the table below:

Baud	States	Bits/Second
250	2	250
250	4	500
250	16	1000
250	32	2000

Notice that at a fixed baud rate of 250, as the number of possible states per symbol increases, the bit rate also rises:

- Two states: Bit rate equals the baud rate (250 bits/sec).
- Four states: Each symbol conveys two bits, double the original bit rate (500 bits/sec).
- Eight states: Now it's three bits per symbol, which further increases the bit rate (750 bits/sec).

With only two possible states per symbol, the bit rate matches the baud rate (250 bits/sec). As the number of states per symbol doubles (4, then 8, then 16), the bit rate doubles, quadruples, and increases eightfold. This exponential rise greatly increases signal transmission efficiency. When they encode multiple bits into each symbol, digital modes can transmit more information within the same bandwidth. The FCC's symbol rate (baud) regulations indirectly govern the maximum achievable bit rates for various modes.

Protocol and Modulation

A protocol outlines the precise guidelines that dictate how digital data is prepared, transmitted, and decoded. It defines aspects such as the specific format of the data, the valid character set (letters, numbers, symbols), and strategies to deal with potential errors during transmission.

Modulation is how digital information is imprinted onto a radio wave. Packet radio frequently uses FM or SSB modulation. The FT8 mode illustrates the combination of protocol and modulation. Its protocol dictates the structure of the data, while it uses a modulation method called 8-FSK. This information is then carried via an SSB transmitter, often in USB mode.

Frequency Shift Keying (FSK)

In FSK, the carrier wave's frequency is not constant but switches between two distinct frequencies to represent the data sent. These two frequencies used in FSK are traditionally designated as "mark" and "space." Mark corresponds to a binary 1 and is usually the higher frequency. Space represents a binary 0 and is often the lower frequency. The difference between the mark and space frequencies is called the shift keying bandwidth. However, the shift (difference) can vary based on the chosen digital mode and application. A common shift for RTTY is 170 Hz.

Two main methods to generate FSK signals are direct FSK and audio FSK.

- **Direct FSK**: In this method, a digital signal from the computer directly controls the transmitter's VFO or synthesizer. This allows precise control over the frequency shifts.

- **Audio Frequency Shift Keying (AFSK)**: This method uses audio tones of varied frequencies to represent the digital data. The computer generates these tones, which are then fed into the microphone input of an SSB or FM transmitter. The transmitter modulates the carrier wave with these audio tones, which essentially "piggybacks" the digital information onto the audio tones. A few precautions must be taken to ensure transfer fidelity. The audio signals transmitted used to represent the data must be free from background noise, interference, or distortion within the audio signal. These confounders can disrupt the receiver's ability to correctly identify the mark and space frequencies and cause errors in the decoded data. Similarly, automatic level control (ALC) and audio compression circuits should be disabled in AFSK transmission because they can distort the audio tones used in AFSK.

Radioteletype (RTTY)

RTTY is one of the oldest digital modes in amateur radio. It uses a unique character coding system and specific transmission techniques to transmit text over the airwaves.

The Baudot Code: At the heart of RTTY lies the Baudot code, which represents each character as a unique five-bit sequence. A start bit and a stop bit are also used to help the receiver station stay synchronized with the transmission. To illustrate, the letter "A" is represented in the code as the five-bit sequence "11000." However, with only 32 possible combinations of five bits, the code cannot represent the entire alphabet, numbers, and punctuation symbols. To overcome this, RTTY uses "LTRS" and "FIGS" codes to switch between two character sets, which doubles the available characters.

RTTY on HF

The most frequently used RTTY speeds on HF are 45, 50, and 75 baud (which correspond to approximately 60, 67, and 100 words per minute). A standard frequency shift of 170 Hz between the mark and space tones is the norm. For successful communication, stations must agree on the baud rate and the frequency shift.

Phase Shift Keying (PSK)

As discussed earlier, PSK is a digital modulation technique where information is encoded through the shift of the phase of a carrier signal. PSK alters the point within the carrier waveform's cycle. Common types within amateur radio include BPSK (binary phase shift keying) and QPSK (quadrature phase shift keying).

BPSK and PSK31: BPSK uses two distinct phase shifts, usually 0 and 180 degrees, to represent digital 1s and 0s. PSK31 is a popular digital mode that uses BPSK modulation. The "31" in PSK31 signifies its approximate symbol rate of 31.25 baud (symbols per second). PSK31 enhances efficiency through the use of Varicode. Varicode is a code system with variable length (that is, it assigns shorter bit sequences to frequently used letters and longer ones to less common characters). This is conceptually similar to Morse code. The number of data bits conveyed in a single PSK31 character fluctuates due to the Varicode system. Use of uppercase letters, which necessitate longer code sequences, will slightly reduce transmission speed.

QPSK and QPSK31: QPSK uses four distinct phase shifts (often 0, 90, 180, and 270 degrees) to represent pairs of bits. This equips QPSK to transmit more data within the same bandwidth compared to BPSK. QPSK31 is a variation on PSK31 that uses QPSK modulation. Due to the manner in which it encodes two bits of data simultaneously into phase shifts, QPSK31 requires stations to agree on the sideband in use. Mismatched sidebands will prevent successful communication.

To improve transmission reliability, QPSK31 incorporates error correction capabilities. This helps the receiving station recover the original data, even if some interference or noise is encountered during transmission. Despite the increased data-carrying potential of QPSK31, it maintains a narrow bandwidth requirement. This bandwidth is comparable to that of BPSK31, and this contributes to its efficiency on crowded HF bands.

Packet-Based Modes and Systems

Packet Basics

In contrast to continuous data streams, packet radio transmits information in discrete, structured units called packets or frames. Each packet contains the following key components:

- **Header**: This initial section carries essential information such as routing directives, control signals to handle the packet, synchronization patterns for the receiver, and data to detect and potentially correct errors.
- **Data (Payload)**: This segment houses the actual information intended for exchange. Data compression techniques are often applied to maximize efficiency.
- **Trailer**: The trailer may contain additional control information, status details, and data for error detection.

Error Detection and Correction

A cyclic redundancy check (CRC) is a common method used to detect errors introduced during transmission. The packet's contents must be calculated, with the result included for comparison at the receiving end.

Two main techniques help ensure accurate data transmission in the presence of noise or interference: forward error correction and automatic repeat request.

- **Forward Error Correction (FEC)**: FEC proactively embeds extra, redundant information into the transmitted data. The receiving station uses this redundancy to identify and correct errors, which reduces the need for retransmissions.

- **Automatic Repeat Request (ARQ)**: ARQ protocols are designed for reliable data transmission between a single sender and receiver. If a detected error occurs, the receiver requests a retransmission of the faulty packet. In ARQ systems, if the receiving station detects errors in a packet, it sends a "NAK" (negative acknowledgment) signal. This signal triggers the transmitting station to resend the problematic packet. This process continues until the data is received without errors.

ARQ and PACTOR

PACTOR, a proprietary digital mode, uses an ARQ protocol for error correction. Consequently, PACTOR connections are inherently restricted to point-to-point communication between two stations. No new stations can join an existing PACTOR exchange.

PACTOR, VARA, and Interference

Although PACTOR and VARA protocols exhibit some tolerance to interference and fading, they are not entirely immune. Signs that excessive interference might disrupt PACTOR or VARA communications include:

- **Repeated Retries**: The systems engage in numerous retransmission attempts or experience frequent timeouts.

- **Delayed Transmission**: Noticeable pauses occur during message transmission.

- **Failed Connections**: Stations may be unable to establish a successful connection.

- **ARQ Response to Interference**: ARQ modes have limits. If attempts to exchange information fail repeatedly due to interference, the connection will eventually be dropped.

Winlink: Email Over Radio

Winlink is a type of packet radio that provides a unique system used to connect amateur radio stations to the internet. It enables radio operators to send and receive standard email messages, even when traditional internet access is unavailable. Amateur Emergency Services often rely on Winlink when traditional infrastructure is compromised. Winlink also provides an essential service for communication from vessels at sea. It is maintained by volunteer amateur radio operators worldwide.

Winlink Infrastructure

Winlink supports both VHF and HF operations, which expands its reach and flexibility. Key components include the servers, modems, and email clients.

- **Servers**: There are remote message servers (RMS) and common message servers (CMS). RMS, also called gateways, bridge the radio network and the internet. CMS links different RMS stations and allows messages to reach their destination across the Winlink network.

- **Modems**: Specialized hardware modems (like PACTOR) or software modems (like VARA) are needed to interface the radio with the Winlink system.

- **Email Messaging Client**: Winlink Express is a specialized email client software that has become the standard choice to manage and compose messages within the Winlink system. Winlink Express is often used together with the VARA software modem. It supports HF (VARA HF) and VHF (VARA FM) operations. VarAC is a derivative of VARA designed for direct peer-to-peer (P2P) chat communication on HF bands, independent of the main Winlink infrastructure.

WSJT Modes

WSJT modes are a suite of digital radio protocols designed to extract extremely weak signals. WSJT stands for "Weak Signal Communication by K1JT." K1JT is the call sign of Joe Taylor, the principal creator of these digital protocols. The core purpose of these digital modes is to decode signals that are extremely faint or buried in noise.

Digital modes like FT8, JT65, JT9, and WSPR, collectively known as the WSJT modes, can decode signals far below the noise floor. They can successfully receive transmissions up to 30 dB weaker than environmental noise and can potentially use one thousand times less power.

JT65

JT65 was initially designed with a difficult application in mind: Earth-Moon-Earth (EME) communication. EME bounces signals off the moon, a distance of almost 240,000 miles. By comparison, the earth's circumference is roughly 25,000 miles. The communication distances underpin the incredible sensitivity JT65 was created to achieve.

FT8

FT8 was developed by Steven Franke and Joe Taylor (who pioneered other WSJT modes) and officially introduced on June 29, 2017. “FT” stands for “Franke-Taylor” in recognition of the system’s creators. “8” refers to the 8-FSK modulation scheme it uses. FT8 stands out due to its extremely narrow bandwidth achieved through eight-tone FSK with a mere 6.25 Hz spacing between tones. In addition, its 15-second transmission cycles enable significantly faster contacts than JT65.

Characteristics of FT8

- **Essential Information Exchange**: FT8 is mostly intended to confirm radio contacts with the minimum necessary information. It usually involves the exchange of call signs, grid locators, and signal reports. This information exchange makes FT8 unsuitable for extended conversations.

- **Synchronized Timing**: FT8 uses strict 15-second transmit/receive cycles, which start at the minute and half-minute marks. Precise computer time synchronization (within one second) is essential.

- **Modulation**: FT8 employs an FSK format, which uses 8-FSK.

- **Weak Signal Performance**: FT8 has an exceptional ability to decode signals buried within noise. Even with very low signal-to-noise ratios, successful communication is possible. A signal report of +3 dB indicates a signal with equivalent strength to the environmental noise within a 2.5 kHz bandwidth.

Good FT8 Practices

- **CQ Responses**: When you reply to a CQ call, choose an unoccupied frequency within the alternate 15-second transmit/receive cycle of the calling station. This minimizes the chance of interference.

- **DX Considerations**: Do not transmit directly on a DX station’s frequency, especially when they are calling other stations. Respect for the DX station helps enable successful contacts for all.

- **Stay Clear**: As band conditions change, monitor activity and proactively adjust your transmit frequency to keep your signal in an unoccupied space.

- **Troubleshoot DX Contacts**: If a desired DX station does not respond, try to transmit on a different clear frequency. This might improve signal reception for the DX station.

WSPR

WSPR (pronounced "whisper") stands for weak signal propagation reporter. Its main function is to investigate how radio signals travel under various conditions and use minimal power. WSPR beacons transmit a compact packet of information that includes the station's call sign, its Maidenhead grid locator, and the transmission power level.

Receiving stations around the world listen for WSPR signals and automatically upload received data to the WSPRnet database. This creates a valuable resource used to analyze propagation paths. WSPR is widely used to gather data on the newly allocated 630-meter and 2200-meter bands. It is also an excellent tool to compare the effectiveness of different antenna setups.

Mesh Networks

Mesh networks provide a flexible and adaptable communication solution. They establish a decentralized network where interconnected devices (nodes) cooperate to relay data. Mesh networks benefit amateur radio in many ways, such as multiple frequency options, hardware resourcefulness, reliability, and the scalability of coverage areas.

- **Multiple Frequency Options**: Mesh networks can operate on amateur bands (900 MHz, 2 GHz, 3 GHz, and 5 GHz) and provide options based on propagation needs and equipment availability.

- **Hardware Resourcefulness**: Standard internet routers can be transformed into mesh network nodes when they are loaded with specialized firmware ("Ham Wi-Fi"). This transformation, often accompanied by legally permitted power increases, makes deployment cost-effective.

- **Expansive Coverage**: The coverage area of a mesh network depends on the number of nodes deployed. They can be used for localized communication at special events or extended to cover entire towns or disaster-affected zones.

- **Reliability and Resilience**: A key strength of mesh networks is their robustness. If a single node becomes unavailable, the network can dynamically reroute data through alternate pathways to ensure the message is delivered.

Mesh networks excel when traditional communication systems may be unreliable during emergencies or large-scale community events. They allow the rapid establishment of high-speed data networks to support essential services.

Receive and Transmit Digital Modes

Each digital mode adheres to established conventions about whether USB or LSB is used. Adherence to these conventions is essential for successful communication.

Troubleshoot Decoding Difficulties

If difficulties arise when you need to decode digital signals, consider the following potential factors:

- **Mark and Space Configuration**: Reversed settings for the mark and space frequencies can hinder successful decoding.
- **Baud Rate Accuracy**: Verify that the selected baud rate precisely matches the digital mode used.
- **Sideband Selection**: An incorrect sideband can invert the relationship between the audio tones and the digital data they represent. Ensure proper sideband selection for the chosen mode.

Bandwidth of Digital Modes

The FCC measures bandwidth occupied by a digital signal in the same manner as any other transmission type. This measurement provides insight into how much spectrum a particular mode uses. This table compares the approximate bandwidths required by various digital modes.

Mode	Approximate Bandwidth (Hz)
WSPR.	6.
FT8.	50.
PSK31.	50.
RTTY.	200.
MFSK16.	300.
JT65.	350.
DominoEX.	524.
Olivia.	1,000.
WINMOR.	1,600.
MT63.	2,000.
PACTOR-III.	2,300.

These bandwidths are approximate and can vary based on the specific symbol rates used for each mode. Digital modes like FT8 and WSPR are exceptionally narrow. This narrowness makes them ideal for crowded bands or weak signal work. Modes such as WINMOR, MT63, and PACTOR-III occupy wider bandwidths.

Bandwidth and Symbol Rate

A fundamental relationship exists between symbol rate (how quickly symbols change) and bandwidth. Faster symbol rates directly translate to a wider bandwidth requirement. Think of bandwidth as ripples on a pond that are caused by a disturbance. The faster the disturbance (akin to a faster symbol rate), the wider the ripples (bandwidth) will spread.

Techniques exist to increase the data rate of a digital mode without the need to expand bandwidth. When more bits are encoded within each symbol, higher data transfer is achievable, though the same symbol rate is maintained.

Transmitter Duty Cycle

The duty cycle refers to the percentage of time a transmitter actively transmits compared to its total operational cycle. Some digital modes, like RTTY and PSK31, operate with a near-continuous duty cycle. That is, they transmit signals for extended periods. ARQ modes also tend to have high-duty cycles, often approaching 100% during transmissions.

High-duty cycles that are used for prolonged durations can exceed a transmitter's average power rating, even if the peak power is within limits. To reduce the risk of damage to equipment, it is often advisable to reduce the transmitter's output power to approximately 50% of its maximum when digital modes with high-duty cycles are operated.

Digital Mode Signal Quality

Like voice modes, digital transmissions can suffer from distortion and splatter if audio levels are too high. Overmodulation often manifests visually on a waterfall display as vertical lines that surround the primary signal. Proper data audio levels are essential for a clean, well-defined digital signal.

ALC and Digital Modes

ALC does not allow excessive signal levels to reach a transmitter's amplifier. This guards radio equipment against potential damage. To do this, ALC reduces the audio level if it becomes too high. ALC can introduce distortion, especially if it aggressively compresses peaks within a signal and reduces the audio level. Digital modes are generally less tolerant of distortion than SSB voice transmissions.

It is recommended that ALC is deactivated when you transmit digital data to minimize the chance of ALC distortion of the transmitted signal. It also helps to reduce the microphone gain level to a point where ALC cannot activate. The transceiver's manual may be of help, as some models (like Elecraft) may provide specific guidance for optimal configuration when digital modes are used.

Interpret Waterfall Displays

A waterfall display provides a visual representation of signals within a selected frequency range. It provides insights that standard spectrum displays often do not. Three important things to pay attention to when you use a waterfall display are highlighted below:

- **Time**: The vertical axis depicts time. Signals scroll slowly downward on the display and provide a visual activity history.
- **Frequency**: The horizontal axis represents frequency. The specific frequencies displayed depend on the configured display window (for example, 14.000 to 14.350 MHz).
- **Signal Strength**: The intensity or color of a signal on the waterfall indicates its strength. Stronger signals generally appear brighter or more saturated in color (often red).

Chapter 9: Antennas and Feed Lines

Antennas act as the invisible bridges between transmitting and receiving devices in our wireless world. They consist of conductive elements that transform electrical signals into radio waves for transmission and vice versa. The orientation of these elements determines the polarization of the radiated wave, which dictates its alignment in space. Horizontal elements generally produce horizontally polarized waves, while vertical elements produce vertically polarized waves. This alignment is important for efficient signal transmission and reception, as antennas with mismatched polarization will experience significant signal attenuation.

Feed Point Impedance

Feed point impedance represents the resistance the antenna presents at the point where it connects to the transmitter or receiver. When an antenna is resonant (has its electrical properties tuned to a specific frequency), its feed point impedance becomes purely resistive. This property ensures optimal power transfer between the antenna and the connected device and minimizes energy loss.

Visualize Antenna Performance with Radiation Patterns

The performance of an antenna is further characterized by its radiation pattern. Imagine a sphere that surrounds the antenna, with the antenna at its center. The radiation pattern is a graphical representation of the signal strength emitted by the antenna in different directions across this sphere. It often consists of two main elements:

- **Lobes**: These are regions within the pattern where the signal strength is strongest. The primary lobe represents the direction in which the antenna preferentially transmits or receives signals.

- **Nulls**: These are points within the pattern where the signal strength is minimal. Nulls occur naturally in the radiation pattern and can be used to minimize interference from unwanted signals.

Types of Antennas and Gain

The world of antennas boasts a diverse range of designs, which each possess unique properties. Discussed below are the most common types relevant to amateur radio transmissions.

- **Isotropic Antenna**: This theoretical concept serves as a reference point that radiates equally in all directions. Although not physically realizable, it provides a benchmark to compare the performance of practical antennas.
- **Omnidirectional Antenna**: This type of antenna, as the name suggests, radiates a uniform signal strength across all horizontal directions. These antennas are ideal for applications where coverage in all directions is necessary, such as cellular base stations.
- **Directional Antenna**: In contrast to omnidirectional antennas, directional antennas concentrate their signal strength in specific directions. This ability, known as gain, is measured in dB relative to a reference antenna (often a dipole or an isotropic source). A higher gain signifies a more focused signal and has advantages such as increased range and reduced interference. Directional antennas find applications in satellite communication, point-to-point radio links, and high-gain Wi-Fi connections.

Front-to-Back Ratio (F/B) and Front-to-Side Ratio (F/S)

These metrics further quantify the directionality of an antenna. The F/B ratio compares the signal strength in the primary direction (front lobe) to the signal strength in the opposite direction (back lobe). A high F/B ratio indicates a well-defined primary lobe and minimal signal leakage in the opposite direction. Similarly, the F/S ratio compares the signal strength in the primary direction to the signal strength at right angles to the main direction (side lobes). A high F/S ratio indicates a clean radiation pattern with minimal signal leakage to the sides.

Skin Effect and Coaxial Cables

At higher frequencies (radio frequency range), a phenomenon known as the skin effect occurs. This effect dictates that current tends to flow along the outer surface of conductors. The skin effect has important implications for coaxial cables, commonly used to connect antennas to transmitters and receivers. At radio frequencies, a coaxial cable effectively behaves as if it has multiple conductors—the outer surface of the inner conductor and the inner and outer surfaces of the shield. This complex behavior can sometimes cause unintended radiation from the feed line in dipole antennas fed with coax, as the shield itself may radiate if not properly grounded or choked.

Dipole Antenna Variations and Vertical Antennas

The **half-wave dipole** is a classic antenna design comprised of two quarter-wavelength elements fed at the center. It has a nominal 73-ohm impedance (a good match for common feed lines) and a broadside radiation pattern that directs its signal perpendicular to the wire. The antenna's height above ground influences its pattern. An antenna that is closer to the ground produces a more omnidirectional pattern, while one with a greater height focuses the signal toward the horizon.

An **end-fed dipole** receives its feed at one end rather than the center. This design requires a transformer to match the antenna's altered impedance to the feed line. End-fed dipoles are often used in situations where support at one end of the antenna is more practical.

The **off-center fed dipole (OCFD)** provides convenient multi-band operation through the selection of an off-center feed point along the antenna elements. This flexibility makes OCFDs popular for radio operators who want to use multiple frequencies without the need to change antennas constantly.

Quarter-wave vertical antennas radiate against a ground plane, which can be Earth or a set of radials that simulate an ideal ground. They have a nominal 36-ohm impedance and often require a matching mechanism for use with standard 50-ohm feed lines. Vertical antennas are susceptible to ground losses, where part of their signal is dissipated as heat in the ground.

Loaded Antennas

When space constraints limit the physical size of an antenna, it can be made to appear electrically longer through the addition of inductance (coils) or capacitance (like top hats or discs). Loaded antennas are often employed in situations where space is limited, such as mobile installations or smaller properties. However, this technique comes with a trade-off. The addition of these reactive components narrows the antenna's usable bandwidth compared to a full-size antenna that operates at the same frequency.

Gain Specifications

Antenna gain, expressed in dB, is a measure of how efficiently an antenna directs signal strength compared to a reference. When the gain is referenced in dBd, the antenna is compared to the performance of a standard dipole antenna. This comparison is grounded in the practical world since dipoles are widely used. On the other hand, dBi references gain relative to a theoretical isotropic radiator that radiates equally in all directions. Since a real isotropic antenna is impossible, dBi is a more idealized reference.

A dipole itself has 2.15 dB more gain than an isotropic radiator. To convert from dBi to dBd, subtract 2.15 dB (a gain of 6 dBi is roughly equivalent to 3.85 dBd). In the same vein, add 2.15 dB to a dBd to convert it to dBi. For example, an antenna rated at 8.15 dBi has an equivalent gain of 6 dBd (8.15 – 2.15 = 6). Knowledge of which specification is used ensures a correct comparison between antennas.

Multiband Antennas

Multiband antennas are designed to function across multiple frequency bands (for example, 10 and 20 meters). These provide an advantage for radio operators who need to work on several frequencies without the need to switch antennas every time. This ability comes with an important trade-off: multiband antennas lack harmonic filters. If a transmitter malfunctions and produces unwanted signals at harmonic frequencies (multiples of the current operating frequency), a multiband antenna is likely to radiate those harmonics as well, which can cause interference to other radio users.

Dipole Antennas

Dipole Basics

A dipole antenna consists of two straight conductors, each of which measures approximately one-quarter of the wavelength of the intended signal, arranged horizontally with a feed point at the center.

Radiation Pattern

Dipoles radiate energy perpendicular (broadside) to their axis. When visualized in free space, they create a figure-eight pattern in the horizontal plane. This pattern means signals are strongest off the sides of the antenna, with minimal radiation directly off the ends.

Resonance, Length, and Examples

The ideal theoretical length of a dipole for resonance is calculated as *492 ÷ frequency (MHz)*. However, to achieve perfect resonance in practice, factors that can affect the electrical length must be considered and adjusted for.

- **Wire Thickness**: Thicker wires have a lower electrical length than thin wires. The difference means a dipole made from thicker wire will need to be slightly shorter to achieve resonance at the desired frequency compared to a dipole made from thinner wire.
- **Environment**: The presence of nearby structures and conductors can alter the current distribution within the antenna to change its electrical length. Nearby metal objects or even the ground itself can influence the resonant frequency. In

some cases, the dipole length might need slight adjustment to compensate for the environment.

For example, you may need to estimate the length (in feet) of a half-wave dipole antenna that will resonate at 28.40 MHz. Use the formula given above [*492 ÷ frequency (in MHz)*] and choose the option closest to your answer. Substitute for frequency in the formula, and you get 492 ÷ 28.40 = 17.324 feet. The answers may be 18 feet, 30 feet, 12 feet, and 75 feet. The closest option to the answer here is 18.

Variations of the Basic Dipole Antenna

The basic dipole design supports several variations to adapt to different constraints and requirements. Examples of these variations include the end-fed half-wave (EFHW) antenna and the inverted-V dipole.

- **End-Fed Half-Wave (EFHW)**: An EFHW antenna is fed at one end rather than the center. This dramatically alters the impedance characteristics. EFHW antennas usually have very high feed-point impedances (often in the thousands of ohms). They almost always require an impedance-matching transformer or network to work effectively with standard 50-ohm coaxial cable.

- **Inverted V**: Here, the dipole elements are bent at the center to form an inverted "V" shape on a central support point. The main advantage of an inverted V is its ability to conserve horizontal space. If you cannot stretch a dipole fully across its ideal length, the inverted V allows you to achieve some height but still fit the antenna into a smaller footprint.

Ground-Plane Antennas

Ground Plane Basics

A ground-plane antenna is essentially half of a dipole antenna. The other half is replaced by a conductive ground plane, which can be a sheet of metal, the body of a car, the earth itself, or a network of wires called radials. Ideal ground-plane antennas have a vertical element that measures a quarter-wavelength of the target frequency and are fed at the base where they connect to the ground plane.

Resonance, Length, and Examples

The ideal length of a ground-plane (vertical) antenna for resonance is estimated by the formula *246 ÷ frequency (MHz)*. Double the formula to estimate the resonant length of a traditional dipole antenna.

Use this formula for the same scenario given above for the dipole antenna: *246 ÷ frequency (in MHz)* = 246 ÷ 28.40 = 8.66 feet. The answers may be 20 feet, 9 feet, 12 feet, and 15 feet. The closest option to the answer here is 9.

Radiation Pattern

Vertically oriented ground planes generate omnidirectional radiation, and signals radiate equally in all directions horizontally. This radiation pattern is valuable for mobile communications where the direction of other stations may constantly change.

Impedance

An ideal ground-plane antenna has a feed-point impedance of approximately 36 ohms. However, this can be adjusted. For instance, you can bring the impedance closer to 50 ohms for a better match to a standard coaxial cable. To do this, angle the radials downward.

Mobile Applications

Effective HF antennas ideally need to be a significant fraction of the wavelength they aim to transmit and receive. For example, a full-size antenna for the 20-meter band is around 33 feet long. A 33-foot antenna on a car is not practical, and this is where the concept of loading becomes useful.

Loading techniques make an antenna appear electrically longer than its physical size. This elongation allows it to resonate at the desired frequency but still fit within the limitations of a vehicle. Some of the most common loading techniques, like loading coils, screwdriver antennas, capacitance hats, and linear loading, are discussed below.

Loading coils involve the addition of a coil (inductor) at some point along the antenna to increase its electrical length. The coil's placement and the number of turns influence how much "extra" length it provides. An increase in the number of turns in the coil increases its inductance, which means the antenna appears electrically longer.

Base loading involves the placement of the loading coil at the base of the antenna and has the most significant effect on the antenna's electrical length. It is the most common method for heavily shortened antennas, like those on vehicles. **Center loading** positions the coil somewhere in the middle of the antenna to provide a moderate degree of electrical length. In **distributed loading**, instead of a single coil, the inductance can be spread along the antenna with multiple smaller coils. Distributed loading can improve bandwidth but might be less practical to construct.

Screwdriver antennas incorporate a variable inductor (coil) within their structure. Through an increase or decrease in the inductance in the antenna system, variable inductors can change the antenna's electrical length. This property allows them to resonate at different frequencies without the need to alter the length of the antenna's whip physically. The benefit of screwdriver antennas is their ability to adjust the resonant frequency from a remote location quickly. This ability is perfect for mobile use, where you might want to change bands or adapt to different propagation conditions without the need to stop to change the antenna.

Capacitance hats require that a disk, radial wires, or some other structure be placed at the tip of the antenna. These appendages add capacitance to the antenna. Similar to coils, capacitance hats also lower the resonant frequency and make the antenna seem electrically longer.

Linear loading involves a portion of the antenna folded back on itself (sometimes in a coil-like shape) to increase the electrical length without the need to extend the physical length.

Corona balls are not used for loading per se. They are mitigation devices used on some mobile antennas, especially those with high voltage at the tip. As radio waves travel, they can ionize air molecules at the antenna tip, especially when high voltages are encountered. This ionization creates a small region of conducting plasma that can cause a faint glow (corona discharge). Corona discharge wastes energy and can introduce noise into the transmission. Corona balls provide a smoother, more rounded surface at the tip of the antenna. This reduces the concentration of the electric field and prevents corona discharge. They usually consist of a metal sphere or multiple radiating elements around the tip.

Effects of Ground Planes on Impedance

The proximity of a ground-plane antenna to its conductive ground plane (the car body, Earth, or radials) significantly impacts its feed-point impedance. The closer the antenna element is to this ground plane, the lower the impedance becomes. This relationship is critical because most radio systems are designed with a 50-ohm impedance standard. It is essential to match the antenna's impedance to this value to minimize signal reflections (SWR) and ensure the maximum amount of power from your transmitter is actually radiated as radio waves.

Several techniques help control impedance in ground-plane antennas. If you raise the antenna even slightly from the ground plane, this decreases the impedance. If you angle radials downward (around a 45-degree angle), this is another effective method to increase impedance closer to the 50-ohm target. In some cases, where placement adjustments are not enough, impedance-matching networks (like transformers) might be needed at the antenna's feed point for a perfect match.

Effects of Ground Planes on Radiation Patterns

Reflections from the ground plane interact with direct radiation in complex ways. This interaction creates areas of reinforced signal (lobes) and areas where the signal cancels out (nulls). The height of the antenna above the ground plane is a major factor when you need to determine the shape of this pattern.

At low heights (below half a wavelength), most of the energy is directed upward. This pattern is excellent for NVIS propagation, where signals go up, then bounce off the ionosphere and come back down within a region. When the antenna is higher (above half a wavelength), complex lobes form at various angles. This can be beneficial for long-distance (DX) communication, as some lobes will be directed at low takeoff angles to reach distant stations.

Polarization

Polarization refers to the orientation of the electric field component of an electromagnetic wave. In simpler terms, it is the direction in which the electric field vibrates as the wave travels. There are three main types of polarization relevant to antennas:

- **Horizontal Polarization**: The electric field vibrates horizontally, perpendicular to the direction the wave travels. Imagine waves that move away from you like ripples on a pond, and the electric field vibrates from side to side.
- **Vertical Polarization**: Here, the electric field vibrates vertically, up and down, relative to the direction the wave travels. The electric field vibrates like a jump rope that is waved up and down as it moves forward.
- **Circular Polarization**: A combination of horizontal and vertical components creates an electric field that rotates. This type is less common for mobile antennas but can be used in specific applications.

Generally, horizontally polarized antennas experience less signal loss from ground reflections. This is because the horizontal electric field component does not interact with the ground plane as strongly as a vertical component. However, there are situations where vertical polarization can be advantageous.

Low-mounted antennas can be used to make long-distance contacts as a result of polarization. When a vertically polarized antenna is mounted close to the ground (below half a wavelength), it can sometimes outperform a horizontally polarized antenna for DX contacts. This augmentation occurs because the ground reflection can reinforce the signal at certain angles.

Yagi Antennas

Directional Antennas and Azimuthal Maps

Antennas distribute radio energy in distinct patterns. Omnidirectional antennas, such as ground planes, radiate energy broadly in horizontal directions, while dipoles radiate energy more directionally in a figure-eight pattern. Directional antennas, like the Yagi, focus energy into a concentrated beam. This provides gain (signal strength increase) in the desired direction and reduces noise and interference from elsewhere.

Precise aim is essential to maximize the benefits of directional antennas. This is achieved through the use of an azimuthal map as a guide. The main lobe of the antenna must also be aimed toward the receiving station. Azimuthal projection maps are invaluable tools for precise aim. When centered on a station's location, these maps display true bearings and distances to all other points on the globe and help to orient an antenna correctly.

Main Lobes and Rotators

A directional antenna's **main lobe** is the main direction in which a directional antenna focuses its energy. Think of it as the antenna's spotlight, where signal transmission and reception are the strongest. Aim the main lobe directly at the station you want to communicate with. This maximizes signal strength to achieve clear, reliable communication.

Rotators are mechanical devices that physically turn your directional antenna to facilitate precise control over its orientation. Control boxes in your station allow easy rotation of the antenna and usually display its current direction in compass degrees. Rotators optimize communication. They position the main lobe toward desired stations and place nulls toward noise sources for improved signal-to-noise ratios.

Core Elements of the Yagi Antenna

Despite its directional power, the Yagi-Uda antenna is built on a few fundamental components. These core elements—the driven element, directors, and reflectors—interact with each other to shape the antenna's radiation pattern and achieve its remarkable performance.

- **Driven Element (DE)**: This resonant dipole is the core of the Yagi design. It is approximately a half-wavelength long, directly connected to the transmitter or receiver, and is where the radio signal enters or exits the antenna system. Think of it as the engine that powers the antenna.
- **Directors**: Multiple directors, shorter than the DE and spaced along the boom in front of it, reinforce the signal in the desired direction. They manipulate the phase of the radio waves and create an effect that amplifies the signal like a series of focused lenses.
- **Reflectors**: One or more reflectors, slightly longer than the DE and located behind it, reflect a portion of the signal energy toward the driven element. This further boosts the signal strength in the forward direction and enhances the antenna's ability to suppress signals that arrive from behind.

Optimize Yagi Design

Certain influential factors must be considered to optimize the gain, front-to-back ratio, and bandwidth of the Yagi antenna. These factors—number of directors, boom length, element diameter, element spacing, and tuning—are discussed below.

Number of Directors

When you add directors in a Yagi design, this progressively concentrates the antenna's signal in a desired direction to boost gain. This is because each director focuses and directs radio waves forward, increasing the gain, while the reflector behind the driven element helps block signals from the rear. However, there is a limit. Too many directors become counterproductive, cause unwanted interactions, and diminish overall gain.

Boom Length

The boom's length significantly impacts how the different elements of the Yagi antenna interact. A longer boom generally allows for more directors to be added, which increases the antenna's gain by concentrating the signal more in a specific direction. However, there is a limit to this effect. Excessively long booms can create complex interactions between the elements and cause phasing problems that decrease the gain instead. To find the ideal boom length, you must maximize the number of directors and avoid negative interactions.

Element Diameter

Element diameter mainly dictates the SWR bandwidth of the Yagi antenna. Bandwidth refers to the range of frequencies over which the antenna operates effectively. Thicker elements create a wider bandwidth, which means the antenna can transmit and receive signals across a broader frequency range without a significant drop-off in performance. This becomes important if you want to use the antenna over a wider band of frequencies.

Element Spacing and Tuning

The precise spacing between the reflector, driven element, and directors, as well as the fine-tuning of their lengths, impacts the antenna's performance. Spacing and tuning are critical to achieve the desired gain, front-to-back ratio, and optimal impedance matching. Varied element spacing affects the way energy is coupled between the elements. It alters the impedance and the radiation pattern (how the signal is distributed in different directions). Fine-tuning the lengths helps resonate each element to the desired frequency.

Impedance Matching for Efficient Power Transfer

Yagi antennas often have a feed point impedance that's higher or lower than the usual 50-ohm impedance of coaxial cables. This mismatch causes signal reflections back into the transmission line, decreases efficiency, increases power loss (shown by a high SWR), and can harm the transmitter. Techniques like the gamma match and beta match are used to correct this impedance mismatch to ensure maximum power is transferred to the antenna for efficient signal transmission.

Gamma Match

The gamma match has the advantages of both mechanical simplicity (the driven element can remain connected to the conductive boom) and a wide range of adjustable impedance values. It functions like a small, parallel transmission line. One conductor is the driven element itself, and the other is a gamma rod connected to the boom. A variable capacitor within the gamma rod allows for the adjustment of its electrical length.

When you move this connection point along the driven element and adjust the capacitor, the gamma match acts as a variable impedance network. This network allows you to "tune" the feed point impedance to match the characteristic impedance of your coaxial cable (usually 50 ohms).

Beta Match (Hairpin)

Beta matches provide a narrower impedance matching range than gamma matches but are generally more resilient against weather conditions due to their enclosed design. The beta match involves a short stub of conductor placed directly across the feed point of the driven element. This stub looks like a hairpin shape, hence the name. The stub acts as an inductor.

The length and position of the beta match stub determine the amount of inductive reactance it creates. This inductance is used to offset any capacitive reactance present at the antenna's feed point. When the inductive and capacitive reactances cancel each other out, you achieve a good impedance match to your transmission line.

Trade-Offs in the Yagi Design

Gain Versus Bandwidth: An increase in the number of directors and a lengthened boom will boost your antenna's gain. However, this comes at the cost of narrower bandwidth.

Front-to-Back Ratio Versus Gain: The front-to-back ratio indicates how much stronger the antenna transmits in the desired direction compared to unwanted directions. Optimize this ratio to minimize interference and noise. To achieve the best front-to-back ratio, you sometimes have to slightly reduce overall gain.

Bandwidth Versus Wind Load: An increase in the diameter of a Yagi's elements improves its frequency range. This wider bandwidth is important for operation across multiple bands or a wide portion of a single band. However, wider elements experience greater wind load, which can increase the risk of damage.

Antenna Modeling Software

Tools like EZNEC and MMANA-GAL are used to design and simulate Yagi antenna performance before construction. These programs facilitate the exploration and optimization of various design parameters. Antenna modeling software provides a valuable resource for interested ham radio enthusiasts.

Loop Antennas

Unlike linear antennas like dipoles, loop antennas form complete circuits. Think of a regular dipole antenna bent into a shape like a square, circle, or triangle. Radio energy can be fed directly into the loop through a small gap or indirectly via a smaller loop that magnetically couples to the main loop. The way current flows around the loop's wire fundamentally determines how the antenna radiates electromagnetic waves. This pattern of current varies based on the loop's size relative to the wavelength of the signal.

Large Loops

Large loop antennas, as the name suggests, have a circumference larger than a single wavelength of the radio signal they operate on. This size difference influences their current flow and signal radiation patterns.

One-Wavelength Loop Antenna

A one-wavelength loop antenna is a closed-circuit antenna where the total circumference of the loop is approximately equal to the operating wavelength of the radio signal. This loop radiates strongest in a direction perpendicular (broadside) to the plane of the loop, unlike dipoles, which radiate strongest along the length of the antenna.

How a large loop antenna is oriented impacts where the signal travels best. Vertical setups are ideal for DX communication. The antenna focuses its signal strength toward the horizon and increases the chance of distant stations being reached. Horizontal configuration is more suited for local and regional communication. The signal strength is strongest in the plane of the loop, which makes it effective for shorter-range contacts.

More Than One Wavelength

As loop size increases beyond a single wavelength, the current distribution within the loop becomes more complex. This complexity causes the formation of multiple lobes (areas of concentrated radiation) and nulls (areas of minimal radiation) in the antenna's radiation pattern. Imagine a bumpy landscape instead of a smooth hill for the signal strength distribution. These multi-wavelength loops tend to radiate more at higher angles, sometimes even straight up, compared to a one-wavelength loop.

The N4KC Skywire Loop Antenna

The N4KC Skywire Loop is a large, non-resonant loop antenna with a circumference of approximately 475 feet. This design allows it to operate effectively across most amateur HF bands. Unlike many antennas that are tuned for specific frequencies, the Skywire Loop functions well over a broad frequency range.

It radiates and receives relatively equally in all directions (omnidirectional) but exhibits a complex radiation pattern with multiple lobes (areas of concentrated signal) and nulls (areas of reduced signal). This means the direction of maximum signal strength will vary based on the operating frequency.

The antenna favors a higher angle of radiation that makes it well-suited for local and regional communication under favorable conditions. The N4KC design uses a 450-ohm ladder line for its feed system, with low loss and ease of construction. However, due to the antenna's inherent impedance characteristics, an antenna tuner must properly match it to the transmitter's standard 50-ohm output.

Quad and Delta Loop Beams

Similar to how a Yagi antenna uses multiple linear elements, quad and delta loop beams employ multiple loop-shaped elements to achieve directionality and gain. These beams usually include:

- **Driven Element**: This is the main radiating element, about one wavelength in circumference for quads. Delta loops use a slightly smaller driven element, and each leg often measures about one-third of a wavelength.
- **Reflector**: This is slightly longer than the driven element (usually around 5% longer) to enhance the signal in the forward direction.
- **Directors**: Multiple directors, slightly (about 5%) shorter than the driven element, further focus the signal into a tighter beam.

Cubical Quads

Invented by Clarence C. Moore, the cubical quad antenna is advantageous in high-power applications or environmentally complex locations. In this design, both the driven element and reflector form square loops. Each side of these loops is approximately one-quarter wavelength long.

Cubical quads have broader bandwidth, a lower radiation angle, potentially shorter boom length, and higher radiation resistance when compared to Yagi antennas of similar gain. For additional gain, cubical quads can be stacked. The feed point location determines the polarization: top or bottom feeding causes horizontal polarization, while side feeding yields vertical polarization.

Delta Loops

Delta loop beams are closely related to cubical quads but use triangular elements instead of square ones. In a two-element delta loop configuration, each leg of the driven element measures approximately one-third of a wavelength, and the reflector is slightly (5%) longer.

Important Comparisons

Quad antennas generally exhibit slightly higher gain than delta loop antennas with the same number of elements, due to the larger surface area enclosed by their square-shaped elements.

A two-element quad or delta loop antenna (made up of a driven element and a reflector) provides similar forward gain (signal strength in the desired direction) as a three-element Yagi antenna.

Yagi antennas generally have a superior front-to-back ratio than quad or delta loops with the same number of elements. This means Yagis are best used to suppress signals and noise that arrive from behind the antenna.

Small Loops

When a loop antenna's circumference shrinks to significantly less than a wavelength (often less than one-third), its behavior changes dramatically due to a relatively uniform current distribution within the loop. This produces unique properties:

- **Sharp Nulls**: Small loops exhibit pronounced nulls perpendicular (broadside) to the plane of the loop.
- **Direction Finding**: These sharp nulls make small loops excellent directional receiving antennas. When you rotate the antenna and pinpoint the direction where a signal disappears (the null), you can locate the signal source.
- **General Reception**: Outside the nulls, small loops receive signals relatively equally from most directions (near-omnidirectional pattern).

Applications of Small Loops

- **Receiving**: Small loops are widely used to receive. Active small loop antennas may include internal amplifiers to improve reception further.
- **Transmitting**: Although less efficient than larger antennas, small loops can be used for transmission when space is severely limited or for temporary setups. They are often vertically oriented.

Magnetic Loops

Larger loops suitable for transmissions are often referred to as magnetic loop antennas. These antennas might incorporate advanced features like automatic tuning mechanisms for ease of use.

Halo Antennas

Imagine that you take a standard half-wavelength dipole antenna and bend it into a circle or square but leave a small gap between the ends. This is how a halo antenna is made. Despite this unusual shape, the halo still radiates its signal much like a loop antenna.

The halo's main radiation pattern focuses outward in a ring parallel to the ground if mounted horizontally. This creates a signal that travels equally in all directions (omnidirectional) horizontally—although it is not truly omnidirectional, as nulls may appear at certain elevation angles. Horizontally mounted halos also produce horizontally polarized waves.

Specialized Antennas

End-Fed Wire Antennas

As discussed earlier, end-fed wire antennas provide a simple and convenient way to get on the air. They feed the antenna at one end instead of the center like a traditional dipole. However, a significant downside of end-fed antennas is that they can exhibit dangerously high RF voltages, especially near the feed point. It is essential to take precautions to prevent RF burns or shocks.

Random Wire Antennas

Similar to end-fed antennas, random wire antennas use a wire length that is not necessarily tuned to a specific frequency. This flexibility can be useful in situations where a resonant antenna is not practical. Like end-feds, they can also have high RF currents concentrated near the feed point, and this increases the risk of shocks.

Stacked Yagi Antennas

When you vertically stack two or more identical Yagi antennas, this provides several benefits. First, it increases the antenna's gain (up to about 3 dB for two stacked Yagis) for a stronger transmitted and received signal. Second, it reduces the vertical beamwidth and focuses the signal into a narrower area. For advanced setups, the radiation pattern of stacked Yagis can be further controlled. When you carefully adjust the phase of the signal that drives each Yagi, the elevation angle of the main beam can be dynamically steered.

Trapped Multiband Dipoles

These use cleverly designed traps (tuned circuits) to allow a single dipole antenna to function on multiple amateur bands. One potential downside is that trap dipoles do not always provide excellent suppression of harmonics.

Log Periodic Antennas

These antennas are known for their wide bandwidth. That is, they function effectively across a broad range of frequencies. However, they provide lower gain than a Yagi of comparable size and are mechanically more complex to build. They also do not inherently suppress harmonics.

Beverage Antennas

Beverage antennas are used to receive signals. They consist of a long and low horizontal wire. Their strength lies in excellent noise rejection, which makes them a good choice when you want to receive weak signals. Note that they have 'negative gain,' meaning they are not suitable for transmitting, although they can be used to transmit in some limited situations with proper matching.

Feed Lines

When there is a mismatch, the power intended for the antenna reflects back. This reflection causes problems like wasted energy and potential damage to equipment. Feed lines form the bridge between a transmitter and an antenna. The relevant concepts about feed lines are discussed below.

Impedance Matching

In an ideal system, all energy from a transmitter seamlessly flows into the antenna for radiation. This requires the antenna's impedance and the feed line's characteristic impedance to be equal. When there's a mismatch, a portion of the signal reflects back toward the transmitter. These reflections produce standing waves on the feed line, a condition measured by SWR.

Standing Wave Ratio: Indicator of Mismatch

SWR reveals the ratio of maximum voltage to minimum voltage along the feed line. A perfect match produces an SWR of 1:1. Higher SWR values indicate a greater degree of mismatch and reflected power. A lossy feed line can mask the true SWR at the antenna, with the SWR measurement at the transmitter being deceptively lower.

To calculate SWR, compare the antenna's load impedance to the feed line's characteristic impedance. Divide the higher value by the lower value. Consider the examples below.

A 50-ohm feed line and a 75-ohm antenna will have a SWR of (75 ÷ 50) 1.5:1. Since the antenna impedance is higher, divide the antenna impedance by the feed line impedance. For a 150-ohm feed line and a 75-ohm antenna, the SWR will be (150:75) 2:1.

High SWR has several negative consequences. Reflections reduce the power that reaches the antenna for radiation. Increased currents due to high SWR on a lossy feed line turn energy into heat and waste power. High SWR can also stress a transmitter's output circuitry and may cause damage. Always aim for the lowest possible SWR to ensure maximum signal strength and safeguard equipment. Choose a low-loss feed line when you use a high SWR.

Antenna Tuners

Antenna tuners (couplers) create a mismatch to compensate and make the transmitter "content." However, they do not resolve the original mismatch at the antenna. The SWR on the feed line between the tuner and the antenna remains unchanged, so line losses are still a concern.

Feed Line Choices

Open-wire feed lines like ladder lines and twin leads generally have lower losses than coaxial cables. Coaxial cables provide ease of use but have inherent signal losses that increase with frequency. Common impedance values for feed lines include 50 ohms and 75 ohms for coax and 300 or 450 ohms for open-wire types.

Chapter 10: Electrical and RF Safety

Although an enjoyable hobby, amateur radio operation carries inherent risks associated with electrical systems and RF energy. This chapter outlines the essential electrical and RF safety considerations, such as proper wiring and grounding techniques, electrical shock prevention, RF exposure limits, and safe outdoor installations.

Electrical Safety

Your ham radio station needs a clearly labeled master on/off switch that is easily accessible in an emergency. Train your family and friends to use the master switch and your home's circuit breaker in case of an electrical accident. Prevention is the most valuable step when it comes to electricity. Listed below are some safety precautions for when you use electrical equipment:

Do not work on electrical equipment that is still powered or 'live' if at all possible. Always double-check with a meter to ensure a device is powered off. Do not work on electrical equipment that is still powered or "live" if at all possible. Always prioritize safety and turn off the power at the source (breaker or fuse box) before you start work. To be sure, use a multimeter to verify that there is no voltage at the points on which you want to work.

Secure transmitters and amplifiers when you work on antennas or feed lines. When you work on antennas or feed lines, ensure that transmitters or amplifiers cannot be accidentally activated. Disconnect the power to these devices and, if you have a partner, implement lockout/tagout procedures with them to prevent accidental power-ups. It might be good to physically disconnect the antenna from the equipment to minimize exposure to high voltages and RF energy.

When you test live equipment, **work with one hand in your pocket**. This simple yet effective technique reduces the risk of electrical shock because it minimizes the chance that you will create a current path through your body. Keep the other hand in your pocket so the current can't flow across your chest if one hand accidentally contacts a live conductor. This break in flow prevents potential heart damage from electricity.

Dress for safety. Wear insulated shoes and remove jewelry when you work with electrical systems. Insulated shoes increase the resistance between you and the ground and reduce the risk of electrical shock. On the other hand, jewelry can conduct electricity and cause burns if it comes into contact with live components. Opt for rubber-soled shoes designed for electrical work and remove all metal objects before you start any electrical tasks.

Discharge capacitors safely. Capacitors can hold dangerous charges long after the power source is disconnected. Always exercise caution when you work with circuits that contain capacitors. Before you touch any component, use a grounding stick (an insulated tool with a conductive tip) to discharge capacitors to a safe level. Follow the manufacturer's instructions or established electrical safety procedures to discharge capacitors and avoid potential shocks.

Secure loose wires and double-check before internal work. Before you work with electrical equipment, take the time to secure any loose wires to prevent accidental short circuits or contact hazards. Remember that components within the equipment may still hold a charge even when the power is off. Use a multimeter to verify that all components are de-energized before you touch them.

Electrical Shock

Electric shocks are a consequence of electric current that passes through the human body. Both AC and DC can inflict severe burns if substantial amounts of currents traverse the body. AC that exceeds a few milliamperes can induce uncontrollable muscle spasms and cause accidental falls or abrupt, uncoordinated bodily movements. The highest current level deemed safe is a mere 50 milliamperes.

The severity of an electrical shock is based on the frequency and path of the electric current flowing through the body. The most dangerous currents traverse the cardiac region and flow from arm to arm or from an arm to a foot. This current pathway disturbs the regular rhythmic contractions of the heart. Household electricity, characterized by low-frequency AC at approximately 60 Hz, is especially dangerous. Its frequency range and ability to penetrate the body can disrupt the heart's conduction system.

In the aftermath of an electrical shock, the heart may enter a chaotic state known as fibrillation. In severe cases, cardiac contractions may cease altogether, a condition called cardiac arrest. Both fibrillation and cardiac arrest can induce immediate unconsciousness, which necessitates prompt intervention and assistance from others to restore vital bodily functions.

Wiring and Safety Grounding

Follow prescribed wiring procedures within residential environments and amateur radio stations. All wiring must adhere to industry-recognized best practices outlined in the National Electrical Code Handbook (NEC).

Moreover, always abide by local building codes so your residence is wired per local requirements. If you violate these regulations, your insurance coverage may be affected.

When faced with doubts about electrical installations, consult with a certified electrician.

Proper wiring conventions should be followed meticulously. Always ensure that the cable or wire selected is appropriately rated to handle the anticipated current load. The standard wire color code dictates that hot (voltage-carrying) wires be black or red color, neutral wires be white, and ground wires be green or bare conductors. Also, always use a circuit tester to double-check the integrity of the installation on completion of any wiring project.

The selection of wire types must be based on their current-carrying capacity ratings, as shown in the table below. The AWG system is a standardized way to measure the diameter of electrical wires. Larger AWG numbers correspond to smaller wire diameters. The current-carrying capacity of a wire depends on its AWG, the type of insulation, and the ambient temperature. The current-carrying capacities in the table are approximate and can vary based on specific conditions. Always consult the NEC for the most accurate and up-to-date information.

Wire Size (AWG)	Diameter (inches)	Diameter (mm)	Approximate Current-Carrying Capacity (Amps)	Common Applications
18.	0.0403	1.024.	10.	Low-power electronics, small appliances.
16.	0.0508	1.291.	13.	Extension cords, lamp cords.
14.	0.0641	1.628.	15.	Household wiring (lights, outlets).
12.	0.0808	2.053.	20.	Household wiring (appliances), larger extension cords.
10.	0.1019.	2.588.	30.	Heavy-duty appliances, air conditioners.
8.	0.1285.	3.264.	40.	Electric stoves, clothes dryers, feeders to subpanels.
6.	0.1620.	4.115.	55.	Service entrance wiring for small homes.

4.	0.2043.	5.189.	70.	Service entrance wiring. for larger homes.

Fault Protections

Protective components safeguard against equipment damage, fire hazards, and shock from equipment malfunctions. These include shock prevention devices designed to detect current imbalances between the input and output conductors. These devices can indicate current that flows into unintended paths, such as through the body. Conversely, power control components protect against excessive current flow. They rapidly disconnect the circuit and prevent fire. Examples are fuses and circuit breakers.

Fuses and Circuit Breakers

As mentioned before, fuses and circuit breakers interrupt excessive current flow to protect equipment from damage. Fuses achieve this via a metal strip that melts and breaks the circuit when overloaded. Circuit breakers use a trip mechanism that also disrupts the circuit but is easily reset. Both fuses and breakers carry specific current ratings that indicate the maximum current they can safely handle. Fuses and circuit breakers also have voltage ratings that specify the maximum voltage they can withstand. Use of fuses or breakers with incorrect ratings can be hazardous as they may fail to interrupt power in an overload or cause dangerous arcing.

Certain situations may call for time-delay fuses. These can tolerate temporary overloads but will ultimately interrupt the circuit if the overload condition persists. A frequently tripped circuit breaker indicates the circuit is overloaded. Solutions include redistribution of loads to other circuits or an increase in the capacity of the wiring and circuit breaker.

Always stick to correctly sized fuses and circuit breakers as stipulated by equipment manufacturers and building codes. Never substitute a fuse or breaker with a higher current rating to bypass an overload issue. That poses a significant safety hazard.

Remember that household power usually arrives as a split 240-volt circuit with two 120-volt hot wires and a neutral wire. Smaller loads use a single hot and neutral, while larger appliances and equipment that require 240 volts use both hot conductors, sometimes with a neutral.

Fuses or circuit breakers are installed exclusively on hot conductors and not on neutral or grounding conductors. Incorrect placement can leave voltages in the circuit even when they appear inactive or severed.

Shock Prevention

Ground Fault Circuit Interrupters (GFCIs)

Ground fault circuit interrupters (GFCIs) are mandated by building codes, especially in areas where water is present, such as kitchens and bathrooms. These devices constantly monitor the current that flows through the hot and neutral wires of a circuit. If a GFCI detects even a tiny imbalance (a few milliamps difference), it rapidly trips and cuts off power.

Why Are GFCIs Important?

The missing current could flow through you instead of back through the neutral wire. This means that you could be in danger of a shock. GFCIs are designed to react immediately to prevent severe or fatal electrical shocks. Therefore, always use GFCI-protected outlets where there is a risk of moisture or direct contact with the ground.

Safety Interlocks

Safety interlocks are vital safeguards often found in ham radio linear amplifiers. They prevent exposure to dangerously high voltages or intense RF fields when cabinets or enclosures are opened for maintenance or repair. They disconnect power or provide a path to discharge high-voltage capacitors.

As a precaution, never bypass an interlock unless absolutely necessary to troubleshoot or explicitly instructed by the manufacturer's manual. Always exercise extreme caution when you work with interlocks disabled. Have another person always present to assist you in case of an accident. Make doubly sure that capacitors are fully discharged. Before you close the enclosure and restore power, double-check that all components and wiring are correctly secured and that no foreign objects are inside.

Generator Safety

Generators can be lifesavers in emergencies or when off the grid. However, it is easy to get too caught up in the convenience and forget essential safety precautions. Many generator-related injuries occur due to fuel mishaps and inadequate ventilation.

Generators release carbon monoxide (CO), an odorless and deadly gas. To prevent CO buildup indoors, always run generators outdoors, well away from windows, doors, and air intakes. Install carbon monoxide alarms in living areas for added protection.

Always shut down the generator before you add fuel, as hot engine parts can ignite spills. Ideally, two people should be present when a generator is refueled—one to handle the fuel, the other to stand ready with a fire extinguisher. Store fuel in approved containers away from the generator.

Before you power your house with a generator, you must physically disconnect your home wiring from the power grid to protect utility workers. This precaution prevents backfeeding. Backfeeding could electrocute linemen at work on a line they think is dead.

Lightning

Lightning strikes may cause fire hazards, damage to electrical equipment, and personal injury. Hence, it requires some safety considerations. Whenever feasible, disconnect all electronic equipment before the onset of a thunderstorm. A comprehensive grounding system is needed for permanently installed outdoor antennas and cabling. All towers, masts, and antenna mounts must have robust grounding systems. These systems should employ thick grounding conductors to direct low-resistance paths to the earth. Use secure bonding techniques (clamps, brazes, or welds) for all ground rods and connections. Consider a halo ground to equalize electrical potential and further enhance safety. Solder is unsuitable due to its potential to melt under the extreme heat of a lightning strike.

Despite meticulous grounding, power surges caused by lightning can still propagate through coaxial cables. Install lightning arrestors designed for radio equipment (such as those manufactured by Polyphasor). Mount the arrestor on a grounded metal plate with a direct connection to the earth. This configuration will shunt harmful high-voltage surges away from sensitive equipment.

Avoid work on antennas or outdoor wires during a thunderstorm. The risk of being struck by lightning is exceptionally high, even with protective measures in place. During severe storms, seek shelter indoors and maintain your distance from electrical appliances. A qualified professional should periodically inspect your lightning protection system to ensure its continued effectiveness.

Radiofrequency Exposure

Power density, specific absorption rate (SAR), time averaging, transmission duty cycles, and the impact of antenna systems are central to the well-being of both operators and the general public. This section covers the factors that influence RF exposure, methods to assess your station's compliance with safety standards, and practical steps to mitigate the risks of RF exposure.

Factors in RF Exposure

The maximum permissible exposure (MPE) is the safety threshold for human RF energy exposure. To calculate MPE, consider the following:

- **Power Density**: Power density is the concentration of RF energy per square meter, much like the brightness of a light. Higher power levels emitted by the antenna produce a higher power density. This increases the potential for hazardous RF exposure. Imagine a spotlight versus a dim light bulb. The spotlight concentrates more light (or, in this case, RF energy) in a smaller space.

- **Frequency**: The human body acts like an antenna but is more "tuned in" to specific frequencies of RF energy. Some frequencies are absorbed more readily than others, like a radio that picks up certain stations better. Therefore, the frequency of the RF radiation is an essential factor to determine the level of exposure.

- **Distance**: The intensity of RF energy diminishes as you move away from the source, just like the heat from a fire decreases the farther you stand from it. In RF exposure, the closer you are to the transmitting antenna, the more RF energy your body will likely absorb.

- **Time**: The longer a person is exposed to RF energy, the greater the amount absorbed, similar to sun exposure. A quick walk in the sun is less harmful than an hour spent sunbathing. The cumulative time spent near a transmitting antenna is critical in RF exposure.

- **Duty Cycle**: This refers to the proportion of time a transmitter actively sends out RF energy. Some transmission modes, like voice (SSB), have intermittent transmissions with gaps in between, while others, like FM or digital modes, transmit continuously. The duty cycle affects the average exposure level, as modes with lower duty cycles have less overall RF output.

RF Exposure Assessment

Metrics like power density and SAR, along with factors like environment, time averaging, and transmission duty cycles, are central to evaluate the risks associated with RF in a station.

Power Density

Power density is the amount of RF energy concentrated within a specific area, measured in milliwatts per square centimeter (mW/cm^2). The amount of RF energy your body absorbs depends on the power density of the field you are in. For example, if an RF field has a power density of 50 mW/cm^2 and 115 square centimeters of your body is exposed, the power density over that area would be 5,750 mW/cm^2 (50×115).

Power density is proportional to the transmitter power of your antenna. That is, higher transmitter output power increases the intensity of RF energy. Power density has an inverse relationship with the distance from an antenna. It is highest near an antenna and in the directions where the antenna has the most gain. When it moves away from an antenna, this reduces the amount of radio frequency energy.

Specific Absorption Rate

The SAR measures how quickly the body absorbs RF energy. It is the most relevant metric to understand radio frequency exposure in ham radio.

The human body absorbs RF energy most efficiently when body parts are similar in dimension to the wavelength of that energy. Different body parts have varied sensitivity to RF energy based on the frequency. In the VHF spectrum (30–300 MHz), the limbs (arms and legs) and torso experience the highest absorption rates (SAR). The head becomes most sensitive as frequencies increase into the UHF range (300 MHz to 3 GHz). The eyes are most vulnerable in the microwave range (greater than 1 GHz). The FCC has established MPE limits to protect amateur radio operators from the hazards of RF energy. These limits are based on the demonstrated dangers of RF energy absorption, quantified by the SAR. MPE limits are tailored to frequencies and account for the variations in how readily the body absorbs RF energy at different points in the spectrum.

Time Averaging and Transmission Duty Cycle

The body's response to RF exposure differs based on whether the exposure is short and intense or spread out over a more extended period. This is why safety regulations use time-averaging when they evaluate RF exposure. This involves the quantification of the cumulative RF exposure levels across a predetermined duration. This method ensures that the total RF energy absorbed over a specific interval remains within safe limits, regardless of whether it happened in bursts or continuously.

Controlled and Uncontrolled Environments

A good grasp of the distinctions between controlled and uncontrolled environments is indispensable to ensure adherence to RF safety protocols and mitigate the effects of excessive RF exposure. People in controlled environments (like ham radio stations or near antennas) know the potential RF exposure and can take measures to minimize it. Uncontrolled environments are accessible to the general public, who may be unaware of the risks.

Due to these differences, the FCC sets stricter RF power density limits for uncontrolled environments, as exposure there is more likely to be unintentional. Hence, the time-averaging rules differ. For **controlled environments**, exposure is averaged over six minutes. In **uncontrolled environments**, exposure is averaged over 30 minutes.

Transmission Duty Cycle

The duty cycle is the fraction of time a transmitter is active, expressed as a ratio or percentage. If transmissions occur in brief intervals, the average RF output over an extended duration will be lower. A reduced duty cycle enables the safe use of higher power levels at the prescribed average exposure thresholds. In amateur radio operations where transmission and reception intervals are approximately equal, the usual operating duty cycle is about 50%.

In addition to the operational duty cycle, there is the emission duty cycle, contingent upon the operational mode. For instance, SSB modulation is characterized by a low emission duty cycle (20%) as transmission occurs exclusively during speech intervals. On the other hand, FM has a 100% emission duty cycle due to a continuous carrier wave.

Antenna System

When you assess RF exposure, you must consider how your antenna and feed line affect the signal's power. High-gain antennas can amplify a signal's average power. This means even a relatively low-power transmitter could generate average power levels that exceed the transmitter's output when paired with a high-gain antenna.

Station Evaluation

Regardless of power level, all fixed amateur stations must assess their potential to cause RF exposure. Mobile and handheld transceivers are not subject to this evaluation due to their temporary location and minimal power output. The FCC provides guidelines about when a routine review is required based on your transmitter's power output (PEP) and operating frequency.

FCC Guidelines for Evaluating RF Exposure.

Band	Power Output (W)
160 meters.	500.
80 meters.	500.
60 meters.	500.
40 meters.	500.
30 meters.	425.
20 meters.	225.
17 meters.	125.
15 meters.	100.
12 meters.	75.
10 meters.	50.
6 meters.	50.
2 meters.	50.
1.25 meters.	50.
70 cm.	70.

33 cm.	150.
23 cm.	200.
13 cm.	250.
SHF (all bands).	250.
EHF (all bands).	250.

This table represents an older standard. Power level-based exemptions are still applicable, but updated guidelines should be followed for accurate compliance.

There are several ways to evaluate RF exposure:

- **Direct Measurement**: This involves the use of specialized RF meters or spectrum analyzers to directly measure the RF field strength in various locations around your antenna. Although this provides the most accurate real-world data, it requires expensive equipment and expertise to use effectively.
- **Computer Modeling**: Specialized software can simulate the RF radiation patterns around your antenna based on factors like antenna type, height, gain, frequency, and transmitter power. This method is more accessible than direct measurement but requires accurate input data and comprehensive knowledge of the software.
- **Tables and Calculators**: Organizations like the ARRL provide simplified tools like tables and calculators to estimate RF exposure levels. These tools usually use basic station parameters (antenna type, height, and power) to estimate a safe distance from the antenna or the power density at a specific location. Although less precise than direct measurement or modeling, these tools provide a convenient initial point for assessment.

Regardless of the method used, the goal is to determine one of two things:

- **Safe Distance:** This is the minimum distance you or others must maintain from your antenna to stay within the safe exposure limits set by regulatory bodies.

- **Power Density at a Location:** This tells you the RF field strength at a specific point around your station. When you compare this value to the MPE limits for that frequency and environment, you can determine if the location is safe for people to occupy.

MPE limits are stricter for uncontrolled environments (areas accessible to the general public) since people there may be unaware of the RF source and unable to take precautions. Whenever you change your transmitter power, antenna type, gain, or height, always re-evaluate RF exposure to ensure you remain compliant with safety guidelines.

When Station Evaluation Exceeds MPE Limits

First, reduce the amount of RF energy your station produces. Consider an antenna with a lower gain to spread the energy over a wider area or reduce your transmitter's power output.

Next, control the potential exposure areas. Be mindful of where your high-gain antennas are directed, especially when you operate with higher power. Avoid areas where people might be present. Use physical barriers, clear signage, or interlocks so people can't enter areas where MPE limits could be exceeded.

Indoor and stealth antennas require attention. Their proximity to people means you must carefully assess MPE limits within your home. Use reliable calculations to ensure safety.

For mobile antenna safety, roof placement is best as it increases the distance from people and makes accidental contact with the antenna less likely. Finally, always use a dummy load when you test transmitters to protect yourself and avoid accidental on-air transmissions during adjustments.

Outdoor Safety

Electrical safety within your station is undoubtedly important. However, numerous elements of a ham radio setup extend beyond the confines of your indoor space. Hence, safety considerations for outdoor installations are equally important as the precautions taken within the indoor environment.

Install Antennas

Poles and transmission lines present constant danger and must be treated with extreme caution. A minimum clearance distance of 10 feet is commonly cited, but a safer practice is to keep your entire antenna system (including the tower or mast) at least 100% of its total height away from any power lines. For instance, a 50-foot antenna setup needs a 75-foot safety margin. This ensures your antenna cannot contact power lines if it falls, is blown down, or collapses.

Maintenance Safety

Elevated work environments, such as towers, roofs, and trees, present inherent risks that necessitate a rigorous approach to safety. To ensure a secure environment, amateur radio operators must prioritize preparation, equipment inspection, and adherence to established safety protocols.

Before you start any elevated work, meticulous preparation is paramount. Equip yourself with suitable safety gear, such as harnesses, helmets, gloves, and robust footwear. Adequate time allocation is essential to accommodate unforeseen delays or obstacles.

A comprehensive safety checklist should be diligently followed before any work is done. This involves a thorough assessment of the tower's structural integrity, guy wires, and support systems. In the case of crank-up towers, you must completely retract and secure them. All climbing gear, such as belts, lanyards, ropes, and load-bearing equipment, should be meticulously inspected to ensure optimal condition.

Climbing gear should be inspected to ensure it is within its service life and appropriately rated for the individual's weight. Carabiners should be securely closed and locked. A safety lanyard, and preferably some redundant lanyards, is mandatory.

Before a person ascends, all AC-powered equipment must be deactivated, disconnected, locked out, and tagged to prevent accidental activation. Transmitters should be deactivated and disconnected from the feed line to mitigate the risk of electrical shock or excessive RF exposure. Weather conditions must be assessed.

During the climb, always maintain a controlled pace and watch your feet. It is of utmost importance to follow manufacturer instructions for all equipment and procedures. This ensures that equipment is used as intended and that all safety precautions are observed, which minimizes the risk of accidents or injuries.

Test 1: Questions

(1) What is the difference in propagation between the 60-meter and 80-meter bands?

(A) 60 meters is more reliable for short-range daytime communication than 80 meters.

(B) 80 meters supports longer-distance communication during the daytime compared to 60 meters.

(C) 60 meters is better for long-range nighttime communication than 80 meters.

(D) 80 meters has better propagation for both short and long distances during the day compared to 60 meters.

(2) Which of the following statements accurately describes the operation of beacon stations in amateur radio?

(A) They broadcast propagation and weather updates only.

(B) Only one can operate on a band within a specific geographic area.

(C) Their frequencies must be coordinated with the Federal Aviation Administration.

(D) They must announce their operating frequencies before transmitting.

(3) On which amateur radio band will one most likely encounter a station using a dipole antenna limited to 100 watts PEP?

(A) 10 meters.

(B) 20 meters.

(C) 60 meters.

(D) 28 MHz.

(4) On which amateur bands is Upper Sideband generally used for voice transmissions?

(A) 160 meters and 80 meters.

(B) 30 meters and 20 meters.

(C) 20 meters and higher frequencies.

(D) 40 meters and below.

(5) In RACES (Radio Amateur Civil Emergency Service), what is the longest duration for regular training exercises and tests without additional permission?

(A) Thirty minutes per week.

(B) One hour per week.

(C) Two hours per week.

(D) One hour per month.

(6) When transmitting CW communication, the operator hears another station's callsign between the dits and dahs. What capability does this demonstrate?

(A) Automatic keyer usage.

(B) Procedural signaling.

(C) Zero-beat matching.

(D) Full break-in (QSK).

(7) Which of these statements about volunteer monitors (VMs) in amateur radio is not true?

(A) Volunteer monitors are licensed amateur radio operators who volunteer their time and expertise.

(B) Volunteer monitors use their equipment to monitor the airwaves for rule violations.

(C) Volunteer monitors can issue fines and penalties to operators who violate regulations.

(D) Volunteer monitors submit reports of their findings to the appropriate authorities for further action.

(8) Which of the statements is true?

(A) Lower sideband is the standard for JT65 when using AFSK.

(B) Winlink is a digital mode exclusively designed for sending images over radio.

(C) FT8 is a high-speed data transfer protocol for real-time video streaming.

(D) Lower sideband is the preferred mode for RTTY transmission via AFSK.

(9) During an HF contest, an operator notices rapid fluctuations in signal strength from a distant station. What is the most likely cause?

(A) Sporadic E layer propagation.

(B) Interference from other stations on the same frequency.

(C) Constructive and destructive interference of skywave signals.

(D) Geomagnetic disturbances.

(10) Which ionospheric layer is responsible for the majority of long-distance HF radio wave propagation during the daytime?

(A) D Layer.

(B) E Layer.

(C) F1 Layer.

(D) F2 Layer.

(11) When operating an HF transceiver, the operator notices a persistent, narrowband signal that causes interference. Which receiver feature will most effectively reduce or eliminate this interference?

(A) Noise blanker.

(B) RF gain control.

(C) Notch filter.

(D) Squelch control.

(12) An amateur radio operator wants to visually inspect the on-off pattern of their Morse code transmission. Which piece of test equipment will they use?

(A) RF power meter.

(B) Dummy load.

(C) SWR meter.

(D) Oscilloscope.

(13) A technician notices interference across a broad range of frequencies on a spectrum analyzer. What is the most likely cause?

(A) A loose antenna connection.

(B) An improperly terminated coaxial cable.

(C) A corroded ground rod.

(D) A sparking electrical contact.

(14) When observing an SSB signal on a spectrum analyzer, what visual change will an operator naturally expect to see after enabling a speech processor?

(A) The signal's bandwidth will narrow.

(B) The signal's carrier frequency will shift.

(C) The peak power of the signal will decrease.

(D) The average power level of the signal will increase.

(15) Which of these modifications will most effectively increase the electrical length of a physically short HF mobile antenna?

(A) Add a solenoid coil near the base.

(B) Install a larger ground plane.

(C) Mount a capacitance hat at the top.

(D) Use a higher-power transceiver.

(16) When is the impedance of an electrical circuit considered purely resistive?

(A) When the circuit has no inductive or capacitive reactance.

(B) When the circuit contains only inductors.

(C) When the circuit contains only capacitors.

(D) When the circuit is resonant at a high frequency.

(17) A radio transmitter's power output is increased from 10 watts to 40 watts. Approximately how many decibels (dB) does this power increase represent?

(A) 3 dB.

(B) 6 dB.

(C) 10 dB.

(D) 20 dB.

(18) An oscilloscope displays a sinusoidal waveform with a peak voltage of five volts. What is the peak-to-peak voltage of this waveform?

(A) 2.5 volts.

(B) 5 volts.

(C) 7.07 volts.

(D) 10 volts.

(19) Brian monitors the voltage of a twelve-volt lead-acid battery under load. To ensure the longest possible lifespan for the battery, which voltage level should prompt him to disconnect the load to prevent excessive discharge?

(A) Below 8.5 volts.

(B) Below 10.5 volts.

(C) 20 volts.

(D) Approaching 12 volts.

(20) A high-frequency amplifier circuit is designed to be fabricated on a single semiconductor chip. What type of integrated circuit is this most likely to be?

(A) ASIC.

(B) FPGA.

(C) MMIC.

(D) SoC.

(21) In a high-voltage power supply, which component helps safely discharge stored energy after the power is turned off?

(A) Rectifier diode.

(B) Filter capacitor.

(C) Bleeder resistor.

(D) Voltage regulator.

(22) Which technique is employed in RF amplifiers to mitigate the risk of undesirable oscillations arising from feedback?

(A) Biasing.

(B) Filtering.

(C) Neutralization.

(D) Impedance matching.

(23) What component selects a superheterodyne receiver's desired intermediate frequency (IF) signal?

(A) Mixer.

(B) Local oscillator.

(C) IF filter.

(D) Detector.

(24) What is "fade margin" in radio communication?

(A) The minimum signal level needed to overcome background noise.

(B) The difference between the received signal strength and the minimum required signal for reliable communication.

(C) The maximum distance a signal can travel without fading.

(D) The strength of the signal at the transmitter before any loss.

(25) What is the function of the local oscillator in a superheterodyne receiver?

(A) Amplify the received radio frequency (RF) signal.

(B) Filter out unwanted frequencies from the RF signal.

(C) Shift the received RF signal to a fixed intermediate frequency (IF).

(D) Demodulate the IF signal to recover the original information.

(26) Which amateur radio band overlaps with frequencies used for Wi-Fi networks in the United States?

(A) 6 meters.

(B) 10 meters.

(C) 2 meters.

(D) 13 centimeters.

(27) What will an antenna system with a high SWR likely experience?

(A) Increased efficiency due to reduced power reflection.

(B) Improved signal quality due to constructive interference.

(C) Higher power loss in the transmission line.

(D) Lower operating temperature at the transmitter.

(28) What potential safety hazard is associated with random-wire HF antennas connected directly to a transmitter?

(A) Increased risk of lightning strikes due to the long wire.

(B) RF burns from high currents flowing back onto station equipment.

(C) Excessive electromagnetic interference to nearby electronic devices.

(D) Overheating of the antenna wire due to high power transmission.

(29) An amateur radio operator wants to build a Yagi antenna for the twenty-meter band. To achieve the widest possible SWR bandwidth, what should they prioritize when selecting materials for the antenna elements?

(A) Choose thin, lightweight aluminum for flexibility.

(B) Use thick copper tubing.

(C) Arrange steel wires very close together.

(D) Utilize any conductive material.

(30) What is the feed point impedance of a half-wave dipole antenna fed at its tips?

(A) 50-75 ohms.

(B) 100-200 ohms.

(C) Several thousand ohms.

(D) Less than a hundred.

(31) Which factors help estimate the degree of exposure to RF a person experiences?

(A) Transmitter power and antenna height.

(B) The transmission's power density, frequency, and duty cycle.

(C) Duration of exposure and the person's body size.

(D) Frequency of the signal and distance from the antenna.

(32) What is the minimum recommended wire gauge for a circuit carrying a maximum of thirty amps, as stipulated by the National Electrical Code?

(A) 14 AWG.

(B) 12 AWG.

(C) 10 AWG.

(D) 8 AWG.

(33) What must a U.S. amateur radio operator do to legally operate in a foreign country under a reciprocal operating agreement?

(A) Apply for a license from the foreign country.

(B) Follow the local regulations of the foreign country.

(C) Use only frequencies assigned in the United States.

(D) Operate only in CW mode.

(34) In which ITU Region is Saudi Arabia located?

(A) Region 1.

(B) Region 2.

(C) Region 3.

(D) Region 4.

(35) Which amateur radio band will likely provide the best long-distance daytime propagation during high solar activity?

(A) 160 meters.

(B) 80 meters.

(C) 40 meters.

(D) 10 meters.

Test 1: Answers and Explanations

(1) (A) 60 meters is more reliable for short-range daytime communication than 80 meters.

The 60-meter band provides more consistent daytime NVIS (Near Vertical Incidence Skywave) propagation, which is ideal for short-range regional communication. The 80-meter band, on the other hand, is more susceptible to signal absorption during the day.

(2) (B) Only one can operate on a band within a specific geographic area.

Beacon stations are coordinated to avoid interference with each other. Thus, only one is usually allowed to operate on a specific band within a given geographic area. These stations transmit their call signs, locations, and power levels, not propagation or weather updates. Their frequencies are not coordinated with the Federal Aviation Administration (FAA). The FAA is concerned with structures that may pose a hazard to air navigation, such as tall towers, not the frequencies used by beacon stations. Beacon stations do not announce their operating frequencies before transmitting. They transmit on a regular schedule, and their frequencies are known to those who monitor them.

(3) (C) 60 meters.

The 60-meter band is allocated to amateur radio operators but is shared with other services. The FCC has designated five channels within this band for amateur use to minimize interference. These channels have a maximum power output of 100 watts ERP (Effective Radiated Power), which is lower than the power allowed on other bands. ERP is the power of a radio signal as it leaves the antenna and accounts for the transmitter's power and the antenna's gain. Dipole antennas are often used to transmit on the 60-meter band due to their size and efficiency at these frequencies. They consist of two conductive elements fed in the center, usually half a wavelength long. They are a common choice for amateur radio operators due to their simplicity and effectiveness. The 10-meter band has a maximum power of 1500 watts PEP. The 20-meter band has a maximum power of 1500 watts PEP. The 10-meter band has a maximum power of 1500 watts PEP and includes the 28 MHz frequency.

(4) (C) 20 meters and higher frequencies.

Upper Sideband is commonly used on frequencies of 20 meters and above for voice transmissions in amateur radio. This includes the 20, 17, 15, 12, and 10-meter bands, as well as the VHF and UHF bands, while Lower Sideband is used on the bands below 20 meters, such as 40, 80, and 160 meters.

(5) (B) One hour per week.

RACES (Radio Amateur Civil Emergency Service) is a program that allows amateur radio operators to provide communication support during emergencies and disasters. Regular training exercises and tests are conducted to ensure that RACES volunteers are well-prepared and proficient in their roles. However, these activities should not disrupt normal amateur radio operations or interfere with potential emergency communications. Therefore, the FCC (Federal Communications Commission) has established a limit of one hour per week for regular RACES training and testing without requiring additional permission. This allows sufficient practice and coordination while minimizing the impact on other amateur radio activities. If a RACES group needs to conduct longer training sessions, they can request a special temporary authorization (STA) from the FCC.

(6) (D) Full break-in (QSK).

CW (Morse code) communication traditionally involves one station transmitting while the other listens. However, with full break-in (QSK), the transmitting station's equipment can detect signals between their dits and dahs (the dots and dashes of Morse code). This allows the receiving station to transmit during those pauses and enables a more conversational communication style where operators can interrupt and respond in real time. Automatic keyer usage helps generate precise Morse code. Procedural signaling involves special characters for specific meanings. Zero-beat matching helps stations receive clear CW signals.

(7) (C) Volunteer monitors can issue fines and penalties to operators who violate regulations.

Volunteer monitors (VMs) are licensed amateur radio operators who generously donate their time and expertise to help maintain the integrity of amateur radio bands. They use their personal radio equipment to listen for signals that may violate FCC rules. These violations include out-of-band transmissions, excessive power levels, or improper operating procedures. However, VMs do not have any enforcement authority. They compile detailed reports of their observations, such as date, time, frequency, call signs, and nature of the violation. These reports are then submitted to the FCC, which investigates the reported violations and takes any necessary action, such as issuing warnings, fines, or even revoking licenses.

(8) (D) Lower sideband is the preferred mode for RTTY transmission via AFSK.

Radioteletype (RTTY) is a form of digital communication that transmits text messages over radio waves. It was initially developed for use with teleprinters. Today, it is commonly used by amateur radio operators for purposes such as casual conversations, contests, and bulletin board systems (BBS). RTTY can be transmitted using different modulation techniques, but one common method is Audio Frequency-Shift Keying (AFSK). In AFSK, the transmitter's frequency is shifted between two tones (mark and space) to represent digital data. The choice between upper sideband (USB) and lower sideband (LSB) depends on convention and frequency.

(9) (C) Constructive and destructive interference of skywave signals.

During an HF contest, rapid fluctuations in signal strength from a distant station are most likely caused by the constructive and destructive interference of skywave signals. Skywave propagation refers to radio waves being refracted (bent) by the ionosphere and returned to Earth, which facilitates long-distance communication. Multiple skywave signals can arrive at the receiver from different paths, each with varying phase and amplitude. When these signals combine, they can reinforce each other (constructive interference) and produce a stronger signal. Conversely, they can cancel each other out (destructive interference) to weaken the signal. This phenomenon, known as fading, is characterized by rapid fluctuations in signal strength and is common during HF contests due to the high number of stations transmitting simultaneously.

(10) (D) F2 Layer.

The ionosphere is a level of Earth's upper atmosphere where solar radiation ionizes atoms to create a layer of free electrons and ions. This layer is divided into several distinct sub-layers: D, E, F1, and F2. Each layer has unique characteristics that affect radio wave propagation. The F2 layer, located approximately 200-400 kilometers above the Earth's surface, is the highest and most densely ionized layer. This high electron density is critical for long-distance HF (High Frequency, 3-30 MHz) radio wave propagation, especially during the daytime when solar radiation is strongest. The F2 layer acts like a mirror that refracts (bends) HF radio waves back toward Earth.

(11) (C) Notch filter.

A notch filter is an electronic filter designed to reject a specific, narrow range of frequencies while allowing others to pass through. This helps eliminate or reduce persistent, narrowband interference, such as those caused by carriers or heterodynes from nearby stations. The notch filter creates a notch or dip in the frequency response curve at the frequency of the interfering signal and reduces its amplitude. A noise blanker is designed to mitigate pulse-type interference caused by electrical devices like automobile ignitions, motors, and power lines. It works by briefly muting the receiver during the short duration of the interference pulse. RF gain control defines the amplification of the radio frequency (RF) signal entering the receiver. Even though reducing the RF gain can help in some situations, it will also attenuate both the interference and the desired signal. Squelch is used to mute the receiver's audio output when no signal is present, primarily to reduce background noise. However, it will not help with an already present and continuous interfering signal.

(12) (D) Oscilloscope.

An oscilloscope is a versatile electronic test instrument that displays voltage signals as waveforms over time. In amateur radio, it can be used to visually inspect the on-off keying pattern of a Morse code transmission. It reveals the precise timing and shape of the dots and dashes. This helps to assess the quality of the code and identify any issues like erratic timing or key clicks. An RF power meter measures the power output of a radio transmitter. Although it can ensure compliance with regulations and optimize antenna performance, an RF power meter does not provide a visual representation of the Morse code pattern. A dummy load is a device that simulates the electrical load of an antenna but dissipates the RF power as heat instead of radiating it. It tests transmitters without an antenna connection and does not provide information about the signal's waveform. An SWR (Standing Wave Ratio) meter measures the degree of mismatch between a transmission line and an antenna. A high SWR indicates reflected power, which can damage the transmitter. SWR meters do not reveal the on-off pattern of a Morse code signal.

(13) (D) A sparking electrical contact.

A sparking electrical contact is the most likely cause of interference across a broad range of frequencies on a spectrum analyzer. Sparking generates broadband electromagnetic interference (EMI), which manifests as noise across a wide spectrum of frequencies. When an electrical connection intermittently breaks and reforms, it causes a visible spark or arc. This can occur in various components, such as switches, relays, motors, or even loose connections. This broadband noise can interfere with various radio services and be easily observed on a spectrum analyzer as a rise in the noise floor across multiple frequencies.

(14) (D) The average power level of the signal will increase.

A speech processor is an audio processing device used in radio communications to manipulate the audio signal before it is transmitted. Its main purpose is to increase the average power of the transmitted signal to make it more intelligible, especially under noisy conditions or at greater distances. In SSB (single sideband) transmissions, the speech processor compresses the dynamic range of the audio and makes the quieter parts louder and the louder parts slightly quieter. This produces a more consistent signal level that can be transmitted at a higher average power without exceeding the transmitter's peak power limit. A spectrum analyzer displays the amplitude (power) of a signal as a function of frequency. Enabling a speech processor will manifest as a general increase in the average power level of the SSB signal across its bandwidth. This is because the processor boosts the weaker parts of the signal and brings them closer to the level of the stronger parts.

(15) (C) Mount a capacitance hat at the top.

A capacitance hat (top hat) is a metallic structure added to the top of a vertical antenna. Capacitance is the ability of a conductor to store electrical charge. An increase in the capacitance of an antenna will increase its electrical length. Electrical length measures how long the antenna appears to the radio wave, which differs from the physical length. A longer electrical length means the antenna can resonate at a lower frequency. Due to vehicle space constraints, mobile antennas are inherently shorter than fixed station antennas. This shorter physical length means they naturally resonate at higher frequencies. The resonant frequency of an antenna is the frequency at which it most efficiently radiates and receives signals. A capacitance hat can artificially extend the antenna's electrical length to make it resonate at lower frequencies within the HF band.

(16) (A) When the circuit has no inductive or capacitive reactance.

Impedance is purely resistive when there is no reactance present in the circuit, meaning it has only resistance and no inductive or capacitive components. This results in current and voltage being in phase, without any phase shift caused by reactance.

(17) (B) 6 dB.

Decibels (dB) are logarithmic units that express the ratio between two power levels. Use the following formula to calculate dB gain: $dB = 10 \times \log_{10} (P2 \div P1)$, where P1 is the initial power, and P2 is the final power. In this case, P1 = 10 watts and P2 = 40 watts.

> $10 \times \log_{10} (40 \div 10) = 10 \times \log_{10} (4) \approx 6$ dB
> 3 dB is a doubling of power: ($10 \times \log_{10}(2) \approx 3$ dB
> 10 dB is a tenfold increase in power: ($10 \times \log_{10}(10) = 10$ dB
> 20 dB is a hundredfold increase in power: ($10 \times \log_{10}(100) = 20$ dB

(18) (D) 10 volts.

Peak voltage (Vp) and peak-to-peak voltage (Vpp) are both measures of waveform amplitude, but they represent different aspects. Peak voltage is the maximum positive or negative deviation of the waveform from its zero reference level. It measures the height of a single peak. On the other hand, peak-to-peak voltage is the total difference between the positive and negative peaks and measures the full height of the waveform from top to bottom. The Vpp is simply double the peak voltage (Vp) for a sinusoidal waveform. Mathematically, $Vpp = 2 \times Vp$. Therefore, if the peak voltage is five volts, the peak-to-peak voltage will be ten volts: $2 \times 5\ V = 10\ V$.

(19) (B) Below 10.5 volts.

Lead-acid batteries are susceptible to damage and reduced lifespan when discharged below a certain threshold. Allowing the voltage to drop below 10.5 volts under load signifies a deep discharge. This can cause irreversible damage to the battery, such as sulfation of the plates, which reduces capacity and shortens the battery's overall life. By disconnecting the load when the voltage drops below 10.5 volts, Brian can ensure the battery does not experience excessive stress and extends its lifespan.

(20) (C) MMIC.

MMIC (Monolithic Microwave Integrated Circuit) is designed for high-frequency applications, especially in the radio frequency (RF) and microwave range. They integrate active devices like transistors, passive components like resistors and capacitors, and even transmission lines onto a single semiconductor chip. This integration allows compact size, improved performance, and reduced parasitics at high frequencies. These properties make MMICs ideal for high-frequency amplifier circuits. ASICs (Application-Specific Integrated Circuits) are designed for specific applications but may not be optimized for high-frequency performance. FPGAs (Field-Programmable Gate Arrays) are reconfigurable digital circuits but are not explicitly tailored for high-frequency analog applications. SoCs (System-on-Chip) integrate various system components, like processors and memory, but may not be optimized for high-frequency analog circuits.

(21) (C) Bleeder resistor.

A bleeder resistor is a component that safely discharges stored energy in a high-voltage power supply after it is turned off. High-voltage power supplies provide electrical power at voltages significantly higher than usual household ones and often range from hundreds to thousands of volts. These power supplies usually contain large filter capacitors that store electrical charge to smooth out the output voltage. Even after the power supply is turned off, these capacitors can retain a dangerous amount of charge. A bleeder resistor provides a path for the stored charge to discharge gradually. This ensures that anyone working on the power supply is not exposed to a potentially lethal electrical shock. Rectifier diodes are used to convert AC voltage to DC voltage. Filter capacitors store energy to smooth out the output voltage of a power supply. Although filter capacitors discharge over time, a bleeder resistor provides a more controlled and faster discharge path. Voltage regulators maintain a stable output voltage and do not actively discharge stored energy.

(22) (C) Neutralization.

Neutralization is a technique used in RF (Radio Frequency) amplifiers to counteract the effects of unwanted feedback, which can cause oscillations. Oscillations are undesirable self-sustained fluctuations in the output of an amplifier, which can distort the signal and even damage the amplifier. Feedback occurs when a portion of the output signal is coupled back to the input. If this feedback is positive (in phase), it can cause the amplifier to become unstable and oscillate. Neutralization introduces a controlled amount of negative feedback equal in magnitude but opposite in phase to the unwanted positive feedback. This cancels out the positive feedback, stabilizes the amplifier, and prevents oscillations. Neutralization is commonly used in RF amplifiers, especially those operating at higher frequencies where parasitic capacitances can cause significant feedback.

(23) (C) IF filter.

In a superheterodyne receiver, the IF (Intermediate Frequency) filter selects the desired IF signal. This type of radio receiver converts the received radio frequency (RF) signal to a fixed intermediate frequency (IF) before demodulation. This conversion is achieved through heterodyning, which mixes the incoming RF signal with a local oscillator signal to produce the IF signal. The IF filter is a band-pass filter tuned to the specific IF frequency and rejects unwanted signals outside this frequency range. This allows the receiver to focus on the desired signal and improve selectivity (the ability to reject unwanted signals). The detector demodulates the IF signal to recover the original information (such as audio or video).

(24) (B) The difference between the received signal strength and the minimum required signal for reliable communication.

Fade margin is a buffer built into the system to ensure reliable communication despite signal fading, which can occur due to factors like weather, obstacles, and interference. It represents the extra signal strength above the minimum needed to maintain a stable connection.

(25) (C) Shift the received RF signal to a fixed intermediate frequency (IF).

In a superheterodyne receiver, the local oscillator (LO) generates a signal at a specific frequency. This signal is then combined with the incoming radio frequency (RF) signal in a mixer. The mixer produces new frequencies, which include the sum and difference of the LO and RF frequencies. The difference frequency is known as the intermediate frequency (IF), which is a fixed frequency chosen for the convenience of further signal processing. An RF amplifier boosts received radio frequency (RF) signals. RF filters remove unwanted frequencies from an RF signal. A demodulator demodulates the IF signal to recover the original information.

(26) (D) 13 centimeters.

The 13-centimeter amateur radio band overlaps with the 2.4 GHz frequency range, which is also used by Wi-Fi networks in the United States. This overlap allows for potential experimentation and use within that frequency range under amateur radio regulations.

(27) (C) Higher power loss in the transmission line.

An antenna system with a high SWR (Standing Wave Ratio) will likely experience higher power loss in the transmission line. SWR is a measure of how well the impedance of an antenna system is matched to the transmission line's characteristic impedance. A perfect match generates an SWR of 1:1, which means all power is transferred to the antenna, and none is reflected. However, when there is a mismatch, some power is reflected, which produces a higher SWR. This reflected power creates standing waves along the transmission line, which alternate high and low voltage and current levels. High SWR reduces the transmitter's efficiency. This generates higher operating temperatures that can cause the transmission line to arc or melt in extreme cases. Standing waves cause destructive interference that creates nodes and antinodes of voltage and current and results in signal distortion and power loss.

(28) (B) RF burns from high currents flowing back onto station equipment.

Random-wire HF antennas can present a safety hazard due to the possibility of RF (Radio Frequency) burns. These antennas are often not resonant at the operating frequency, which can produce high SWR (Standing Wave Ratio). High SWR means a significant portion of the transmitted power is reflected toward the transmitter. This reflected power can cause high RF voltages and currents to appear on the outer shield of coaxial cables and other station equipment. Touching these energized surfaces can cause RF burns similar to those caused by heat.

(29) (B) Use thick copper tubing.

To achieve the widest possible SWR (Standing Wave Ratio) bandwidth in a Yagi antenna, the builder should prioritize using thick copper tubing for the antenna elements. Thicker elements have lower resistance, are less affected by the skin effect, and are more mechanically stable. Thicker conductors have lower electrical resistance, which reduces losses and produces a more efficient antenna. This helps to maintain a better match over a wider range of frequencies. Skin effect is the tendency of high-frequency currents to concentrate near the surface of a conductor. It is less pronounced in thicker conductors. This further reduces losses and contributes to a broader bandwidth. Thicker elements are more rigid and less prone to bending or deformation due to wind or other environmental factors. This stability ensures that the antenna maintains its designed shape and electrical characteristics over a wider range of frequencies.

(30) (C) Several thousand ohms.

A half-wave dipole antenna fed at its tips presents a very high feed point impedance, usually in the range of several thousand ohms. Feed point impedance is seen at the point where the transmission line connects to the antenna. It is vital for efficient power transfer and to minimize signal reflections. When fed at its center, a half-wave dipole has maximum current at the feed point and minimum voltage. This results in a low impedance of around seventy-five ohms (depending on the surrounding environment). However, when fed at its tips, the current is minimum, and the voltage is maximum. According to Ohm's law, this produces a high impedance ($Z = V/I$), where Z is impedance, V is voltage, and I is current.

(31) (B) The transmission's power density, frequency, and duty cycle.

The degree of exposure to RF (Radio Frequency) a person experiences depends on the transmission's power density, frequency, and duty cycle. Power density is the amount of RF energy per unit area, usually expressed in milliwatts per square centimeter (mW/cm^2). Higher power density means more RF energy is present. The frequency of the RF signal determines how the body absorbs the energy, such as penetration depths and amounts. The duty cycle is the percentage of time that a transmitter is actively transmitting. A higher duty cycle means longer exposure to RF energy. Although other factors like transmitter power and antenna height can influence the power density, they are not direct determinants of RF exposure. The duration of exposure and body size are also relevant, but the main factors are the transmission's power density, frequency, and duty cycle.

(32) (C) 10 AWG.

The National Electrical Code (NEC) is a set of standards for the safe installation of electrical wiring and equipment. According to this code, the minimum recommended wire gauge for a circuit carrying a maximum of thirty amps is 10 AWG (American Wire Gauge). The AWG system is a standardized wire gauge classification used in North America for round, solid, nonferrous, electrically conducting wire diameters. The larger the AWG number, the smaller the wire diameter.

(33) (B) Follow the local regulations of the foreign country.

When operating under a reciprocal agreement, U.S. amateur radio operators must follow the regulations and rules of the foreign country they are visiting. The reciprocal operating agreement allows operation without obtaining a separate license but requires adherence to that country's amateur radio rules.

(34) (A) Region 1.

The International Telecommunication Union (ITU) divides the world into three regions for regulatory purposes. Different areas have varying needs and usage patterns for radio frequencies. Dividing the world into regions allows customized regulations and allocations that cater to those specific requirements. This ensures efficient use of the limited spectrum resource while minimizing interference between different services and countries. Saudi Arabia falls under Region 1, which encompasses Europe, Africa, the Middle East, and parts of Central Asia. Region 2 covers the Americas, and Region 3 includes Asia and the Pacific.

(35) (D) 10 meters.

During periods of high solar activity, the sun emits more radiation, such as ultraviolet (UV) and X-ray radiation. This increased radiation ionizes the Earth's ionosphere, especially the F2 layer. The increased ionization density in this layer allows higher-frequency signals to be reflected, which improves long-distance propagation on bands like the ten-meter band (28-29.7 MHz). The lower frequency bands (160, 80, and 40 meters) usually rely on different ionosphere layers (D and E) for propagation. These bands can still provide long-distance communication, especially during nighttime. However, they are less effective during the day when the D layer absorption is higher. The eighty-meter and 160-meter bands are mainly used for nighttime long-distance communication and are not ideal for daytime propagation due to D-layer absorption. Forty meters can sometimes provide daytime long-distance communication but is not as consistently reliable as the ten-meter band during high solar activity.

Test 2: Questions

(1) What statement best describes the restrictions on 30 meters for General Class licensees?

(A) Both phone and image transmission are allowed.

(B) Only phone operation is allowed.

(C) Only image transmission is allowed.

(D) Neither phone nor image transmission is allowed.

(2) Which of the following statements regarding antenna structure height and the requirement to notify the FAA and register with the FCC is true?

(A) Antennas fifty feet above ground must be registered with the FCC.

(B) Operators must notify the FAA about antennas exceeding 100 feet above ground.

(C) Any antenna structure more than 200 feet above ground must be registered with the FCC.

(D) An antenna structure must be at least 250 feet above the ground before the FAA is notified.

(3) If a station on the 10-meter band transmits data at the maximum allowable symbol rate, what is the transmission speed?

(A) 300 baud.

(B) 1200 baud.

(C) 2400 baud.

(D) 9600 baud.

(4) When operating on the 80-meter band, which sideband mode is usually preferred for voice communication?

(A) Upper sideband (USB).

(B) Lower sideband (LSB).

(C) Double sideband (DSB).

(D) Amplitude modulation (AM).

(5) Which licensing assessment is an individual certified as a volunteer examiner (VE) with the highest-level amateur radio license authorized to oversee?

(A) Exclusively the Amateur Extra examination.

(B) Only the General and Technician examinations.

(C) Solely the Technician examination.

(D) All three tiers of amateur radio licensing examinations.

(6) Why do scatter signals often experience distortion?

(A) Weak signal strength.

(B) Multipath interference.

(C) Long distance traveled.

(D) Atmospheric absorption.

(7) In amateur radio, which scenario has the highest priority for frequency usage?

(A) An amateur radio operator attempts to establish a new contact.

(B) A group of operators are engaged in a pre-arranged net discussion.

(C) An emergency where a station must transmit distress signals.

(D) A group of operators participate in a radio contest.

(8) During a CW exchange, one operator sends the Q signal QRS. How should the other operator adjust their transmission in response?

(A) Decrease the code speed.

(B) Increase the transmit power.

(C) Switch to a different frequency.

(D) Slow down and send each character twice.

(9) An amateur radio operator plans a long-distance HF communication. Which map type will be most useful to determine the shortest path between two points on the Earth's surface?

(A) Topographic map.

(B) Mercator projection map.

(C) Azimuthal map.

(D) Political map.

(10) What is PACTOR used for in amateur radio?

(A) Voice communication.

(B) Slow-scan television.

(C) Digital data transmission.

(D) High-speed CW communication.

(11) What will a radio signal that is transmitted higher than the critical frequency experience?

(A) Total internal reflection within the ionosphere.

(B) Absorption by the D layer.

(C) Penetration of the ionosphere and escape into space.

(D) Refraction back to Earth at a steeper angle.

(12) Which scenarios usually cause the highest Maximum Usable Frequency (MUF) for a given path?

(A) Winter, nighttime, low solar activity.

(B) Summer, midday, high solar activity.

(C) Equinox, twilight, moderate solar activity.

(D) Any time of year, nighttime, high solar activity.

(13) Which HF frequencies are generally most affected by Sudden Ionospheric Disturbances (SIDs)?

(A) Lower HF frequencies.

(B) Higher HF frequencies.

(C) All HF frequencies are affected equally.

(D) VHF and UHF frequencies.

(14) Which appendage is designed to reduce high voltage discharge from the tip of an HF mobile antenna?

(A) Capacitance hat.

(B) Gamma match.

(C) Corona ball.

(D) Loading coil.

(15) An operator listens to an SSB transmission and notices the audio sounds distorted and splattery. What is the most likely cause?

(A) The transmitter is operating at reduced power.

(B) The coaxial cable is too short.

(C) The operator is using an excessively long microphone cable.

(D) The transmitter's speech processor is overdriving the signal.

(16) What is the main characteristic of a musical tone in terms of its waveform?

(A) It has a random waveform with multiple frequencies.

(B) It consists of a single frequency with a smooth waveform.

(C) It has a varying frequency with a jagged waveform.

(D) It includes multiple frequencies and is non-repetitive.

(17) To examine the amplitude variations of a radio frequency signal, where should the technician connect the attenuated output of the transmitter?

(A) To the horizontal input of an oscilloscope.

(B) To a dummy load.

(C) To the vertical input of an oscilloscope.

(D) To an antenna analyzer.

(18) What is the main function of passband tuning in a ham radio receiver?

(A) To adjust the receiver's sensitivity.

(B) To increase the audio volume of received signals.

(C) To shift and narrow the receiver's IF passband to reduce interference.

(D) To automatically tune to the strongest nearby signal.

(19) A transmitter produces a peak voltage of 282.8 volts across a 75-ohm load. What is the peak envelope power (PEP) output?

(A) 533 watts.

(B) 2000 watts.

(C) 320 watts.

(D) 426 watts.

(20) An electric heater operates at 220 volts and draws 5 amps of current. Calculate the power consumption of the heater.

(A) 44 watts.

(B) 110 watts.

(C) 550 watts.

(D) 1100 watts.

(21) Which statement accurately describes the relationship between frequency and reactance?

(A) Inductive reactance decreases with increasing frequency, while capacitive reactance increases.

(B) Both inductive and capacitive reactance increase with increasing frequency.

(C) Both inductive and capacitive reactance decrease with increasing frequency.

(D) Inductive reactance increases with increasing frequency, while capacitive reactance decreases.

(22) An operator is designing a battery-powered digital device and needs to prioritize power efficiency. Which logic family will be the most suitable choice for the integrated circuits?

(A) ECL.

(B) TTL.

(C) CMOS.

(D) NMOS.

(23) A power supply designer needs to filter out ripple voltage after rectification. Which type of capacitor will be the best choice due to its ability to store a large amount of charge in a small package?

(A) Ceramic capacitor.

(B) Film capacitor.

(C) Mica capacitor.

(D) Electrolytic capacitor.

(24) What device combines a carrier signal with an information signal to produce an output with two symmetrical sidebands?

(A) Product detector.

(B) Frequency multiplier.

(C) Phase modulator.

(D) Balanced modulator.

(25) Which statement regarding amplifier classes is incorrect?

(A) Class C amplifiers generally have the highest efficiency among the listed classes.

(B) Class A amplifiers prioritize linearity over efficiency and have the lowest efficiency.

(C) Class AB amplifiers are a compromise between Class C's efficiency and Class A's linearity.

(D) Class B amplifiers are more efficient than Class C amplifiers due to their push-pull configuration.

(26) Which statement regarding rectifiers is not true?

(A) A full-wave rectifier has a center-tapped transformer and two diodes.

(B) A half-wave rectifier uses only one diode in its circuit configuration.

(C) A full-wave rectifier converts 360 degrees of the DC cycle to AC.

(D) The pulsating DC output of an unfiltered full-wave rectifier connected to a resistor is at double the frequency of its AC input.

(27) Which statement about WSPR is not true?

(A) It is a low-power mode.

(B) It requires a special license.

(C) It is a digital mode.

(D) It is used for propagation analysis.

(28) What is image response in a radio receiver?

(A) A desired signal appearing at the receiver's output.

(B) An interfering signal appearing at double the intermediate frequency (IF).

(C) A process that converts a received signal to a lower frequency.

(D) A receiver's ability to reject unwanted signals.

(29) Amplitude Modulation (AM) involves varying the strength of a carrier wave. Which statement about AM is correct?

(A) Flat-topping occurs when audio input levels are too low.

(B) Excessive audio levels can cause distortion and increased bandwidth.

(C) The modulation envelope is formed by connecting the minimum values of the AM signal.

(D) Overmodulation causes reduced bandwidth and improved signal quality.

(30) What is the radiation pattern of a halo antenna mounted horizontally?

(A) Omnidirectional in the horizontal plane.

(B) Directional with a narrow beamwidth.

(C) Figure-eight pattern with nulls at the ends.

(D) Omnidirectional in horizontal and vertical planes.

(31) An antenna manufacturer specifies the gain of a Yagi antenna as 12 dBi. What is the equivalent gain of this antenna if expressed in dBd?

(A) 14.15 dBd.

(B) 9.85 dBd.

(C) 12 dBd.

(D) 10 dBd.

(32) How can the feed point impedance of a quarter-wave ground-plane vertical antenna be adjusted to approximately 50 ohms?

(A) Increase the height of the vertical element.

(B) Shorten the radial wires.

(C) Angle the radial wires downward.

(D) Add a loading coil to the vertical element.

(33) Which scenario will reflect the least power at the antenna feed point?

(A) A 50-ohm coaxial cable connected to a 75-ohm dipole antenna.

(B) A 75-ohm coaxial cable connected to a 50-ohm dipole antenna.

(C) A 50-ohm coaxial cable connected to a 50-ohm dipole antenna.

(D) A 75-ohm coaxial cable connected to a 75-ohm dipole antenna.

(34) A ground fault circuit interrupter (GFCI) is designed to protect against what?

(A) Overloading of electrical circuits.

(B) Electrical shocks caused by current leaking to the ground.

(C) Voltage surges from lightning strikes.

(D) Radio frequency interference.

(35) An amateur radio operator wants to confirm that their station's RF emissions are within permissible limits. How can they assess their station's compliance with RF safety regulations?

(A) Refer to the power output specifications of their transmitter.

(B) Consult a table of safe exposure distances for their antenna type.

(C) Use an S-strength meter to estimate emission strength.

(D) Calculate with modeling software and Bulletin 65.

Test 2: Answers and Explanations

(1) (D) Neither phone nor image transmission is allowed.

The 30-meter band has restrictions for General Class licensees. Although they can operate on this band, General Class hams are limited to specific modes such as CW, RTTY (radioteletype), and data. Phone (voice) and image transmissions, such as SSTV (Slow Scan Television), are prohibited on this band for General Class operators. This is due to the band's primary allocation to fixed and mobile services, which require different modulation techniques.

(2) (C) Any antenna structure more than 200 feet above ground must be registered with the FCC.

The Federal Aviation Administration (FAA) requires that any antenna structure exceeding 200 feet above ground level be registered with the FCC. This ensures that the structure does not pose a hazard to air navigation.

(3) (B) 1200 baud.

Baud rate refers to the number of signal events or symbols transmitted per second in a digital communication system. In amateur radio, it represents the speed at which data is transmitted. The maximum symbol rate allowed on the 10-meter band is 1200 bauds.

(4) (B) Lower sideband (LSB).

The 80-meter band (3.500–4.000 MHz) is popular for long-distance communication, especially during nighttime when the ionosphere reflects radio waves more efficiently. This allows signals to travel long distances by bouncing between the Earth and the ionosphere. The ionosphere is a layer of charged particles in the atmosphere. LSB, like USB, is a type of single sideband modulation. It is the preferred mode for voice communication on frequencies below 10 MHz, such as the 80-meter band. This convention was established in the early days of SSB to avoid confusion and ensure that operators on a given frequency were using the same sideband.

(5) (D) All three tiers of amateur radio licensing examinations.

A volunteer examiner (VE) with the highest-level amateur radio license (Amateur Extra) is authorized to oversee all three amateur radio licensing examination tiers: Technician, General, and Amateur Extra. This is because they have demonstrated the highest level of knowledge and proficiency in amateur radio theory and regulations.

(6) (B) Multipath interference.

Scatter propagation involves radio waves bouncing off various irregularities in the atmosphere. This creates multiple paths for the signal to reach the receiver. These paths have different lengths, which causes the signals to arrive at slightly different times. When these out-of-sync signals combine at the receiver, they interfere with each other and produce distortion. This is similar to how echoes in a large room can distort sound. Other factors, like weak signal strength and atmospheric absorption, can affect the overall quality of a scatter signal. However, they are not the main explanation for scatter mode signal distortions.

(7) (C) An emergency where a station must transmit distress signals.

The radio spectrum is a limited resource, and regulations and priorities govern its use. Emergency situations have the highest priority for frequency usage in any radio communication scenario, which includes amateur radio. This priority is based on the fundamental principle of protecting life and safety. Reliable communication is vital for coordinating rescue efforts in an emergency such as a natural disaster or a major accident. It can assist those in need and disseminate critical information. Therefore, any station transmitting a distress signal or engaged in emergency communication has the absolute right of way on the frequency. Regardless of their activity, other stations must immediately yield the emergency communication frequency and refrain from transmitting until the situation is resolved.

(8) (A) Decrease the code speed.

In CW communication, code speed refers to the rate at which characters (dits and dahs) are sent per minute, often measured in words per minute (WPM). The Q signal QRS is a request to send more slowly. If an operator sends this signal, they are having difficulty understanding the transmission at the current speed, likely due to interference or unfamiliarity with the sending style. Decreasing the code speed improves the clarity and readability of the transmission for the receiving operator. QRS can also be used in the question format to mean, "Shall I send more slowly?" The following are other Q signals.

- QRO: "Shall I increase power?" or "Increase power."
- QSY: "Shall I change to transmission on another frequency?" or "Change to transmission on another frequency."

(9) (C) Azimuthal map.

When planning long-distance HF (high frequency) communication, a ham operator must determine the shortest path between two points on the Earth's surface. HF signals often travel through the ionosphere, a layer of charged particles in the upper atmosphere. The ionosphere can refract the signals and allow them to travel great distances. The shortest path between two points on a sphere (like the Earth) is a great circle route, and an azimuthal map shows these great circles as straight lines. Topographic maps focus on land features like elevation, terrain, and water bodies. Mercator maps are commonly used for navigation but distort distances, especially near the poles. This makes them unsuitable for determining the shortest path for HF communication. Political maps show boundaries between countries, states, and other political divisions. They are irrelevant for understanding radio wave propagation.

(10) (C) Digital data transmission.

PACTOR is a digital mode commonly used in amateur radio for reliable data transmission, particularly for email and text messaging over HF bands. It combines features of packet radio and error-correcting protocols to transmit data efficiently over long distances, even in poor signal conditions.

(11) (C) Penetration of the ionosphere and escape into space.

The critical frequency is the highest frequency that can be refracted back to Earth by a particular ionosphere layer at a given incidence angle. When a radio wave encounters the ionosphere, it is either refracted back to Earth or passes through it. If the wave's frequency is higher than the critical frequency for the layer it encounters, the ionosphere will not bend it enough to return it to Earth. Instead, the wave will travel through the ionosphere and eventually escape into space. Total internal reflection occurs when a wave strikes a boundary between two media at an angle greater than the critical angle. Although radio waves can experience some degree of reflection within the ionosphere, total internal reflection is not the primary mechanism for long-distance HF propagation. The D layer primarily absorbs lower-frequency HF signals, not those above the critical frequency.

(12) (B) Summer, midday, high solar activity.

The Maximum Usable Frequency (MUF) is the highest frequency that can be used for skywave propagation between two points at a given time. Several factors influence the MUF. The ionosphere is denser during the summer due to increased solar radiation. This causes higher MUFs. Midday is when the sun reaches its highest point in the sky. This produces maximum ionization and the highest MUF. High solar activity, characterized by increased sunspots and solar flares, further enhances ionization in the ionosphere and produces even higher MUFs. Winter, nighttime, and low solar activity cause lower ionization levels in the ionosphere and, invariably, lower MUFs.

(13) (A) Lower HF frequencies.

Sudden Ionospheric Disturbances (SIDs), triggered by solar flares, unleash intense UV and X-ray radiation that reaches Earth at the speed of light. Solar flares reach the Earth in roughly eight minutes. These flares increase the ionization of the ionosphere's D Region (lowest layer). The D Region absorbs lower HF radio waves. The heightened ionization amplifies its absorption. This absorption causes a sudden decrease or complete loss of signal strength, a phenomenon known as a shortwave fadeout. Although SIDs can impact the entire HF spectrum, lower frequencies are most affected due to their stronger interaction with the increased electron density in the D Region.

(14) (C) Corona ball.

A corona ball is a metal sphere or similarly shaped object placed at the tip of an HF mobile antenna. When transmitting with high power on HF frequencies, the tip of a whip antenna can develop a high voltage potential due to its sharp point. This can cause corona discharge, where the air around the tip ionizes and creates a visible glow or electrical arc. Corona discharge can cause signal distortion, reduce the antenna's efficiency, and potentially interfere with other electronic equipment. A corona ball redistributes the electrical charge at the antenna's tip. It reduces the voltage concentration and prevents corona discharge. The corona ball's larger surface area helps dissipate the charge more evenly and prevent arcs from forming.

(15) (D) The transmitter's speech processor is overdriving the signal.

Overdriving a speech processor occurs when the input audio level to the processor is too high. This can happen if the microphone gain is set too high or the speaker talks too loudly or closely to the microphone. When a speech processor is overdriven, it cannot effectively compress the audio signal's dynamic range. Instead, it starts to clip the peaks of the waveform and introduces distortion that manifests as a splattery sound on the receiving end. Splatter spreads the signal's energy into adjacent frequencies and causes interference with other stations. Operating at reduced power will produce a weaker signal but will not cause the specific type of distortion associated with overdriving a speech processor. The length of the coaxial or microphone cable should not directly cause a distorted or spattered signal unless the cable is damaged or of poor quality.

(16) (B) It consists of a single frequency with a smooth waveform.

A musical tone is defined by a steady, continuous waveform with a single, fundamental frequency. This creates a smooth, repetitive waveform, distinguishing it from noise or complex sounds that contain multiple frequencies or random variations.

(17) (C) To the vertical input of an oscilloscope.

An oscilloscope's vertical input is designed to display the voltage level of a signal. This makes it ideal for examining the amplitude variations of an RF signal. The attenuated output of a transmitter, a reduced-power version of the RF signal, can be safely connected to the vertical input without damaging the oscilloscope. The horizontal input controls the time base, which determines how the signal is displayed across the screen horizontally. Although the time base is essential for viewing the signal's waveform, it does not directly display the amplitude variations. A dummy load is used to absorb RF power for testing transmitters. It does not provide any information about the signal's characteristics. An antenna analyzer measures various properties of an antenna, such as impedance and resonant frequency. Though it can provide some information about the transmitted signal, it is not the best tool to examine its amplitude variations.

(18) (C) To shift and narrow the receiver's IF passband to reduce interference.

Passband tuning allows the operator to shift the receiver's intermediate frequency (IF) passband without changing the tuned frequency. This helps reduce interference from signals on adjacent frequencies by narrowing or repositioning the passband, thereby improving signal clarity and selectivity.

(19) (A) 533 watts.

PEP is the maximum power the transmitter can deliver during a transmission cycle. Use the following formula to calculate PEP: $PEP = Vrms^2 \div R$, where PEP is the peak envelope power in watts (W), Vrms is the root-mean-square voltage in volts (V), and R is the resistance (antenna impedance) in ohms (Ω). In the question, Vp (peak voltage) = 282.8 V and R (antenna impedance) = 75 Ω. Since the formula requires RMS voltage, we must convert the peak voltage to RMS voltage. The formula for this conversion is $Vrms = Vp \times 0.707$.

Substitute the values.
$282.8\ V \times 0.707 = 199.9396$ Vrms.

Use the above value in the PEP formula.
$199.9396\ V^2 \div 75\ \Omega$
$39975.84364816\ V^2 \div 75\ \Omega$
533.01124 W
PEP≈ 533 W

(20) (D) 1100 watts.

Power (P) is the rate at which electrical energy is transferred or consumed. To calculate the power in an electrical circuit, use the formula $P = I \times E$, where P is power, E is voltage, and I is current. In this case, the electric heater operates at 220 volts and draws 5 amps of current.

Substitute the values.
$5\ A \times 220\ V = 1100$ watts

Voltage (E) is the electrical potential difference between two points in a circuit, measured in volts. It can be thought of as the pressure that pushes the electric current through the circuit. Current (I) is the flow of electric charge through a circuit, measured in amperes (A). It can be thought of as the amount of electricity flowing.

(21) (D) Inductive reactance increases with increasing frequency, while capacitive reactance decreases.

Inductive reactance is the opposition to the flow of alternating current (AC) caused by an inductor. An inductor is a passive electronic component that stores energy in a magnetic field when current flows through it. As the frequency of the AC signal increases, the inductor's magnetic field changes more rapidly. These rapid fluxes induce a larger opposing voltage, thus increasing the inductive reactance. Inductive reactance (X_L) is directly proportional to both the inductance (L) and the frequency (f) of the AC signal. The formula for inductive reactance is $X_L = 2\pi fL$. As frequency increases, X_L also increases. On the other hand, capacitive reactance is the opposition to the flow of AC caused by a capacitor. A capacitor is a passive electronic component that stores energy in an electric field. As the frequency of the AC signal increases, the capacitor has less time to charge and discharge. This reduces its opposition to current flow, thus decreasing the capacitive reactance. Capacitive reactance (X_C) is inversely proportional to both the capacitance (C) and the frequency (f) of the AC signal. The formula for capacitive reactance is $X_C = 1 \div (2\pi fC)$. As frequency increases, X_C decreases.

(22) (C) CMOS.

CMOS (Complementary Metal-Oxide-Semiconductor) logic is known for its low power consumption due to its near-zero static power dissipation. This means it consumes minimal power when not actively switching states, which makes it ideal for battery-powered devices where power efficiency is important. ECL (Emitter-Coupled Logic) is known for its high speed, but its high power consumption makes it unsuitable for battery-powered devices. TTLs (Transistor-Transistor Logics) consume more power than CMOS and are less power-efficient. NMOS (N-channel Metal-Oxide-Semiconductor) is a type of transistor technology but not a complete logic family. CMOS, which uses NMOS and PMOS transistors, is generally more power-efficient.

(23) (D) Electrolytic capacitor.

Electrolytic capacitors are the best choice to filter out ripple voltage after rectification due to their high capacitance values and relatively small package size. This high capacitance allows them to store a large amount of charge. This property helps to smooth out the voltage fluctuations (ripple) after AC voltage is rectified into DC voltage. Ceramic capacitors are often used in high-frequency applications due to their low equivalent series resistance (ESR) and inductance. However, their capacitance values are generally too low for effective ripple filtering in power supplies. Film capacitors are known for their stability and low losses. They also have relatively lower capacitance values than electrolytic capacitors, which makes them less effective at storing the large charges needed for ripple filtering. Mica capacitors are known for their stability and ability to handle high voltages. They are not designed for high capacitance values and are thus unsuitable for ripple filtering in power supplies.

(24) (D) Balanced modulator.

A balanced modulator is a device that combines a carrier signal with an information (modulating) signal to produce an output with two symmetrical sidebands. In amplitude modulation (AM), the amplitude of the carrier signal is varied in proportion to the information signal. The resulting output spectrum contains the original carrier frequency plus two sidebands: the upper sideband (USB), which is the sum of the carrier and information frequencies, and the lower sideband (LSB), which is the difference between the carrier and information frequencies. A balanced modulator suppresses the carrier signal and leaves only the two sidebands in the output. This is useful in applications where the carrier signal is unnecessary for demodulation, such as single-sideband (SSB) transmission. A product detector is used to demodulate AM or SSB signals. It multiplies the received signal with a locally generated carrier to recover the original information signal. A frequency multiplier is a circuit that increases the frequency of an input signal by an integer multiple. A phase modulator varies the phase of the carrier signal in proportion to the information signal to produce phase modulation (PM).

(25) (D) Class B amplifiers are more efficient than Class C amplifiers due to their push-pull configuration.

Class C amplifiers are generally more efficient than Class B amplifiers. Class A amplifiers have the lowest efficiency because the active device (transistor) always conducts current, even when there is no input signal. This continuous conduction causes significant power dissipation as heat. However, Class A amplifiers are known for their excellent linearity, which means the output signal is a faithful reproduction of the input signal. Class B amplifiers use two active devices in a push-pull configuration, where each device conducts for only half of the input signal cycle. This reduces the power dissipation compared to Class A amplifiers and results in higher efficiency. However, Class B amplifiers suffer from crossover distortion, where the output signal is distorted around the zero-crossing points.

(26) (C) A full-wave rectifier converts 360 degrees of the DC cycle to AC.

Rectifiers are electrical devices that convert alternating current (AC), which periodically reverses direction, to direct current (DC), which flows in only one direction. They come in two main types: full-wave and half-wave rectifiers. Full-wave rectifiers convert an AC input's positive and negative half-cycles (360 degrees) into pulsating DC. Pulsating DC is a type of DC voltage that varies in amplitude but does not change polarity. The frequency of the pulsating DC output from a full-wave rectifier is twice that of the AC input. This is because each half-cycle of the AC input is converted into a pulse of DC output. A full-wave rectifier configuration uses a center-tapped transformer and two diodes to achieve this conversion. The center-tapped transformer provides two out-of-phase AC voltages to the diodes to rectify both half-cycles. On the other hand, half-wave rectifiers use only one diode. Consequently, only one half-cycle (180 degrees) of the AC input is rectified. The other half-cycle is blocked. This produces a pulsating DC output with the same frequency as the input AC but with gaps where the negative half-cycle would have been.

(27) (B) It requires a special license.

WSPR (Weak Signal Propagation Reporter) is a digital mode amateur radio operators use to investigate radio wave propagation. It is designed for low-power transmission to test the limits of signal propagation. WSPR beacons are usually transmitted at power levels often less than 1 watt and are received by stations worldwide using specialized software. The received data is uploaded to a central database, where it can be analyzed to study propagation patterns. WSPR does not require a special license beyond the standard amateur radio license for the frequency band. It uses a specific digital encoding scheme for transmitting and receiving information.

(28) (B) An interfering signal appearing at double the intermediate frequency (IF).

Image response refers to an undesired signal that interferes with the desired signal in a superheterodyne receiver. It occurs when an incoming signal at an undesired frequency produces the same IF when mixed with the local oscillator signal. This interfering signal is called the image frequency. It is located at a frequency twice the IF away from the desired signal frequency. For example, if the desired signal is 100 MHz and the IF is 10 MHz, the image frequency will be 120 MHz.

(29) (B) Excessive audio levels can cause distortion and increased bandwidth.

In Amplitude Modulation (AM), the strength (amplitude) of a carrier wave is varied in proportion to the information signal (usually audio). When audio input levels are too high, the resulting AM signal can exceed the carrier wave's amplitude limits and cause distortion. Distortion manifests as clipping or flattening of the waveform peaks that cause the signal to sound harsh and garbled. Overmodulation can also increase the bandwidth of the AM signal and interfere with adjacent channels. Flat-topping is a specific type of distortion caused by excessive audio input levels. The modulation envelope is formed by connecting the maximum values of the AM signal. It represents the shape of the original information signal. Overmodulation increases bandwidth due to the generation of harmonics and degrades signal quality due to distortion.

(30) (A) Omnidirectional in the horizontal plane.

When mounted horizontally, a halo antenna exhibits an omnidirectional radiation pattern in the horizontal plane. An omnidirectional antenna radiates radio waves equally in all directions within a plane, in this case, the horizontal plane. The halo antenna's circular shape allows this even distribution of radiated energy. However, its radiation pattern is not omnidirectional in the vertical plane. It has a doughnut-shaped pattern, with a null (area of minimal radiation) directly above and below the antenna. Directional antennas focus their energy in a specific direction, with a narrow beamwidth. A figure-eight pattern is characteristic of dipole antennas.

(31) (B) 9.85 dBd.

The equivalent gain of a 12 dBi Yagi antenna in dBd is 9.85 dBd. Antenna gain measures an antenna's ability to direct or concentrate radio frequency (RF) energy in a particular direction compared to a reference antenna. It is expressed in decibels (dB), a logarithmic unit that compares two power levels or ratios. Two common reference antennas are used to express gain: dBi (decibels relative to an isotropic radiator) and dBd (decibels relative to a dipole antenna). An isotropic radiator is a theoretical antenna that radiates power equally in all directions and is used as a baseline for comparison. Gain expressed in dBi indicates how much better an antenna focuses energy than this theoretical omnidirectional antenna. A dipole antenna is a half-wavelength antenna with a specific radiation pattern. Gain expressed in dBd compares an antenna's performance to that of a dipole in the same direction. The relationship between dBi and dBd is straightforward: dBd = dBi - 2.15 dB. This is because a half-wave dipole has a gain of approximately 2.15 dBi relative to an isotropic radiator. In this question, the Yagi antenna's gain is given as 12 dBi.

> Substitute the values.
> 12 dBi - 2.15 dB = 9.85 dBd.

(32) (C) Angle the radial wires downward.

Angling the radial wires downward on a quarter-wave ground-plane vertical antenna is a common method to adjust the feed point impedance to approximately 50 ohms. The feed point impedance is the impedance seen by the transmission line at the point it connects to the antenna. Radial wires are horizontal or downward-sloping wires that extend outward from the base of the vertical element and work like a ground plane. Angling the radials downward decreases the antenna's radiation resistance, which lowers the feed point impedance. The ideal angle depends on the specific antenna design and operating frequency, but a common starting point is around 45 degrees.

(33) (C) A 50-ohm coaxial cable connected to a 50-ohm dipole antenna.

The least amount of reflected power occurs when the characteristic impedance of the coaxial cable matches the antenna's impedance. Characteristic impedance is the inherent resistance a transmission line presents to the flow of high-frequency electrical energy. Common coaxial cable impedances are 50 ohms and 75 ohms. A dipole antenna is a basic type of antenna that usually has an impedance of around 75 ohms in free space. However, the impedance of a dipole can be modified by adjusting its length or using a matching network. In this scenario, the 50-ohm coaxial cable connected to the 50-ohm dipole antenna is the only one with a perfect match between the cable and antenna impedance. This connection will reflect power the least and transfer the most power to the antenna.

(34) (B) Electrical shocks caused by current leaking to the ground.

A Ground Fault Circuit Interrupter (GFCI) is a safety device that quickly shuts off electrical power during a ground fault. A ground fault occurs when electricity leaks from a circuit and flows to the ground, potentially through a person. GFCIs are designed to detect even small imbalances in current flow between the hot and neutral wires of a circuit. If a ground fault occurs, the GFCI will trip and interrupt the flow of electricity within milliseconds and prevent serious or fatal electric shock.

(35) (D) Calculate with modeling software and Bulletin 65.

To assess compliance with RF safety regulations, an amateur radio operator should calculate the RF exposure levels using modeling software and refer to the FCC's Office of Engineering and Technology (OET) Bulletin 65. OET Bulletin 65 provides guidelines and formulas to evaluate RF exposure and ensure compliance with FCC regulations. Modeling software can help simulate the RF environment around the antenna and estimate the power density at various locations.

Test 3: Questions

(1) What term describes the cumulative RF exposure averaged over a specified period?

(A) Power density.

(B) Specific Absorption Rate (SAR).

(C) Time averaging.

(D) Duty cycle.

(2) What safety precaution is most critical before climbing a tower with electrical equipment installed?

(A) Inform the local power company of climbing plans.

(B) Ensure all power to the tower is disconnected and locked out/tagged out.

(C) Ground the tower's base with a temporary grounding rod.

(D) Wear rubber-soled shoes to insulate from electrical shock.

(3) What Standing Wave Ratio results from connecting a 60-ohm feed line to a 120-ohm resistive load?

(A) 2:1.

(B) 1:2.

(C) 3:1.

(D) 1:3.

(4) In which direction does a quarter-wave ground-plane vertical antenna radiate most of its energy?

(A) Vertically upward.

(B) In a figure-eight pattern.

(C) Toward the base.

(D) Equally in all horizontal directions.

(5) An amateur radio operator wants to increase the range and directivity of their Yagi antenna. Which modification will be most effective?

(A) Shorten the boom.

(B) Add more elements.

(C) Decrease the element spacing.

(D) Decrease the element diameter.

(6) How do antenna traps work?

(A) They act as filters that block unwanted frequencies.

(B) They change the antenna's electrical length for different frequencies.

(C) They amplify the signal on specific bands.

(D) They adjust the antenna's impedance for optimal matching.

(7) In a communication system, which property of a carrier wave is altered in Phase Modulation (PM)?

(A) Amplitude.

(B) Frequency.

(C) Phase angle.

(D) Wavelength.

(8) Which statement below correctly describes heterodyning?

(A) To amplify a radio frequency (RF) signal.

(B) To filter out unwanted frequencies from an RF signal.

(C) To combine two signals of different frequencies to produce sum and difference frequencies.

(D) To demodulate an RF signal to recover the original information.

(9) Which of the abbreviations below indicates a request for packet retransmission in an Automatic Repeat reQuest (ARQ) system?

(A) ACK.

(B) NAK.

(C) SYN.

(D) FIN.

(10) In an NPN junction transistor, which region is connected to a positive voltage in order to allow current to flow?

(A) Emitter.

(B) Collector.

(C) Base.

(D) Substrate.

(11) A logic gate produces a high output only when both inputs are high. What type of gate is this?

(A) OR gate.

(B) NOR gate.

(C) AND gate.

(D) NAND gate.

(12) In a single sideband (SSB) receiver, what component is used to demodulate the received signal by mixing it with a locally generated carrier?

(A) Balanced modulator.

(B) Phase detector.

(C) Frequency discriminator.

(D) Product detector.

(13) A radio enthusiast measures the voltage across a diode in a forward-biased circuit. They observe a voltage drop of approximately 0.3 V. Which type of diode is most likely used in the circuit?

(A) Silicon.

(B) Germanium.

(C) Zener.

(D) Schottky.

(14) Which statement about ferrite core toroidal inductors is not true?

(A) They can be designed for specific inductance characteristics.

(B) Their properties can be adjusted by varying the ferrite material.

(C) They are unsuitable for high-frequency applications.

(D) They provide good magnetic field containment.

(15) What is reactance in an AC circuit?

(A) Opposition to current flow caused by resistance.

(B) Resistance to current flow caused by inductors and capacitors.

(C) Total opposition to current flow, including both resistance and impedance.

(D) Voltage increase caused by capacitors and inductors.

(16) A 5-ohm resistor is connected across a 9-volt battery. What is the power dissipated by the resistor?

(A) 1.8 watts.

(B) 16.2 watts.

(C) 45 watts.

(D) 81 watts.

(17) A transformer is designed to step down the voltage from 120 V to 24 V in a transformer with 600 turns in its primary winding. How many turns does the secondary winding have, and by what principle does the voltage appear in the secondary winding?

(A) 120, mutual inductance.

(B) 300, electromagnetic induction.

(C) 120, ferromagnetic induction.

(D) 300, Lenz's law.

(18) What undesirable effect can overuse of the noise reduction feature cause on a receiver?

(A) Frequency instability.

(B) Increased receiver sensitivity.

(C) Signal distortion.

(D) Reduced power consumption.

(19) A ham operator wants to examine the detailed shape of a radio frequency signal and its peak voltage. Which combination of tools will be most appropriate?

(A) Signal generator and spectrum analyzer.

(B) Oscilloscope and digital multimeter.

(C) Antenna analyzer and directional wattmeter.

(D) Field strength meter and dummy load.

(20) Which of the symptoms below occurs when a ground loop is created in an audio system?

(A) High SWR readings.

(B) Intermittent static noise.

(C) A humming sound in the audio.

(D) Reduced transmit power.

(21) An operator compares two signals on the receiver's S meter. The first signal measures S9. The second signal is 100 times more powerful than the first. Which statement regarding the second signal is correct?

(A) The S meter will display S11 for the second signal.

(B) The S meter will display S9+20 dB for the second signal.

(C) The S meter will display S109 for the second signal.

(D) The second signal will be too strong to register on the S meter.

(22) Which component in a solar power system prevents the battery from discharging back through the solar panel during periods of low or no sunlight?

(A) Inverter.

(B) Charge controller.

(C) Blocking diode.

(D) Voltage regulator.

(23) What statement accurately describes the skip zone in radio propagation?

(A) The area where signals are strongest.

(B) It is found above the F2 layer.

(C) Scatter propagation can bring radio signals to skip zones.

(D) The signals generated in skip zones are always of the best quality.

(24) To achieve the longest possible skip distance on the 10-meter band during a period of high solar activity, an operator should select a frequency:

(A) Just above the critical frequency.

(B) As close to the MUF as possible.

(C) Midway between the LUF and MUF.

(D) As close to the LUF as possible.

(25) Why does the F2 layer enable the longest skip distances for HF radio waves?

(A) It is the highest layer in the ionosphere.

(B) It is the most ionized layer.

(C) It has a higher electron density gradient.

(D) It is less affected by solar radiation.

(26) Which systems allow amateur radio operators to send and receive emails when internet access is limited or unavailable?

(A) Common Message Server (CMS).

(B) PACTOR.

(C) Winlink.

(D) Gateway.

(27) An operator tries to initiate a contact on the HF bands. Which calling sequence is considered standard practice?

(A) CQ CQ CQ DE [Callsign] K.

(B) [Callsign] DE [Callsign] K.

(C) CQ DE [Callsign] QRZ?.

(D) CQ DX DE [Callsign] K.

(28) A radio enthusiast operates on an amateur radio frequency when a distress signal interrupts. Which scenario assists the distressed station the least?

(A) The operator acknowledges the distress signal and determines the nature of the emergency.

(B) The operator immediately ceases their transmission and listens for further instructions from the distressed station.

(C) The operator contacts the local authorities or emergency services to report the distress call.

(D) The operator reduces their transmitter power to minimize potential interference with the distress signal.

(29) Which Q signal will an operator send to inquire if a frequency is currently occupied?

(A) QRN.

(B) QSL.

(C) QRS.

(D) QRL.

(30) Which of the statements below is not an advantage of using single sideband (SSB) over amplitude modulation (AM)?

(A) Requires more complex equipment.

(B) Occupies less bandwidth.

(C) Increased power efficiency.

(D) Improved signal-to-noise ratio.

(31) An amateur radio operator notices that their transmissions cause interference to a nearby television station. What is their immediate responsibility?

(A) Cease transmission immediately and investigate the cause of the interference.

(B) Adjust their antenna to a different direction.

(C) Reduce their transmitter power.

(D) Transmitting on frequencies outside the amateur radio bands.

(32) If an operator holds a General Class license and has passed the Extra Class exam but has not received the new license, can they operate on 15 meters using Amateur Extra mode privileges?

(A) Yes, with a Certificate of Successful Completion of Examination (CSCE).

(B) Only after receiving the physical Amateur Extra license.

(C) No, 15 meters is not an Amateur Extra band.

(D) Yes, but only with reduced power.

(33) Which regulatory body sets the power limits for amateur radio operations in the United States, and what is the standard unit of measurement for these limits?

(A) Federal Communications Commission (FCC), watts RMS.

(B) National Telecommunications and Information Administration (NTIA), watts RMS.

(C) Federal Communications Commission (FCC), peak envelope power (PEP).

(D) National Telecommunications and Information Administration (NTIA), peak envelope power (PEP).

(34) Which statement about the regulation of ham antennas by local authorities is true?

(A) Local governments can sanction ham operators when antenna height exceeds 200 feet.

(B) State governments can remove antenna structures near airports or sensitive areas.

(C) The regulations must be minimal and achieve a valid local objective.

(D) Local authorities can regulate at any time as long as they notify the FCC.

(35) Which statement most accurately describes the 60-meter band's usage in amateur radio?

(A) It is a wideband that allows free frequency selection.

(B) It is exclusively for Morse code communication.

(C) It is shared with other services and thus channelized.

(D) It is only usable during daylight hours due to propagation.

Test 3: Answers and Explanations

(1) (C) Time averaging.

Time averaging describes the cumulative RF exposure averaged over a specified period. RF exposure limits are usually defined in terms of time-averaged values to account for the fact that exposure can vary over time. For example, a transmitter with a low duty cycle (transmitting only a small percentage of the time) can have higher peak power levels while still meeting the time-averaged exposure limits. Power density is the amount of RF energy per unit area. Specific Absorption Rate (SAR) is the rate at which RF energy is absorbed by the body, expressed in watts per kilogram (W/kg). It is not a cumulative measure. The duty cycle is the percentage of time a transmitter is on, but it does not directly represent the cumulative RF exposure.

(2) (B) Ensure all power to the tower is disconnected and locked out/tagged out.

The most critical safety precaution before climbing a tower with electrical equipment installed is to ensure all power to the tower is disconnected and locked out/tagged out. Lockout/tagout (LOTO) is a safety procedure that involves physically locking or tagging electrical switches or circuit breakers in the off position to prevent accidental re-energization while work is being done. This is essential to prevent electrical shocks, burns, or falls from energized equipment.

(3) (A) 2:1.

The Standing Wave Ratio (SWR) measures how well the impedance of an antenna system is matched to the impedance of the transmission line. It is calculated as the ratio of the maximum voltage to the minimum voltage along the transmission line. Equivalently, it is the ratio of the higher impedance to the lower impedance. For a 60-ohm feed line and a 120-ohm antenna, the SWR will be (120 ÷ 60) 2:1. Since the load impedance (120 ohms) is higher, we divide the load impedance by the feedline impedance.

(4) (D) Equally in all horizontal directions.

The vertical element of a quarter-wave ground plane antenna acts as the main radiator that emits radio waves perpendicular to its length. The ground plane (usually consisting of multiple radial wires) acts as a virtual ground. It reflects the radio waves emitted downward by the vertical element to create an image of the antenna below the ground. This reflection combines with direct radiation to produce an omnidirectional pattern in the horizontal plane. Although a small amount of energy might radiate vertically, a quarter-wave ground-plane vertical antenna radiates most of its energy horizontally.

This means it radiates equally in all directions within the horizontal plane (azimuthal). The figure-eight pattern is that of a dipole antenna.

(5) (B) Add more elements.

Operators can add more elements (especially directors) to a Yagi antenna to increase gain, directivity, and front-to-back ratio. Each director acts as a passive amplifier that re-radiates the received signal with a slight phase shift. When these re-radiated signals combine constructively with the signal from the driven element, they produce a stronger signal in the forward direction. Adding more directors increases this effect and produces higher gain. As more directors are added, the antenna's radiation pattern becomes narrower and more focused, which increases the directivity. This allows the antenna to transmit or receive signals over longer distances in the desired direction. More directors also improve the antenna's front-to-back ratio. A higher front-to-back ratio means the antenna is better at rejecting signals coming from behind.

(6) (B) They change the electrical length of the antenna for different frequencies.

Antenna traps are parallel resonant circuits inserted into an antenna element. A resonant circuit is an electrical circuit that oscillates at its natural resonant frequency when excited by an external source. The resonant frequency of a parallel LC circuit (inductor and capacitor in parallel) is determined by the inductor and capacitor values. When a signal's frequency matches the resonant frequency of the trap, the trap presents a high impedance, effectively shortening the antenna for that frequency. Conversely, for frequencies outside the trap's resonant frequency, the trap presents a low impedance. This low impedance enables the signal to pass through, which lengthens the antenna. Placing traps at strategic locations along the antenna can make the antenna resonant at multiple frequencies (making it a multi-band antenna).

(7) (C) Phase angle.

Phase Modulation (PM) is a modulation scheme where the phase angle of the carrier wave is varied in proportion to the information signal. The phase angle refers to the position of a point in time on a waveform cycle. Changing the phase angle can encode the information onto the carrier wave. Varying the amplitude of the carrier wave generates Amplitude Modulation (AM). Varying the carrier wave frequency produces Frequency Modulation (FM). Wavelength is inversely related to frequency, so changing the wavelength is equivalent to changing the frequency (producing FM).

(8) (C) To combine two signals of different frequencies to produce sum and difference frequencies.

Heterodyning is a signal processing technique where two signals of different frequencies are combined in a non-linear device, such as a mixer or a multiplier. The output of the mixer contains new frequencies, which are the sum and difference of the original frequencies. Heterodyning is used in radio receivers to shift the frequency of the received radio frequency (RF) signal to a fixed intermediate frequency (IF). This is done by mixing the RF signal with a local oscillator (LO) signal. The frequency difference between the RF and LO signals becomes the IF, which is easier to amplify and filter. Radiofrequency (RF) signal amplification is done by RF amplifiers, not heterodynes. Filtering out unwanted frequencies from an RF signal is the function of an RF filter. Heterodyning is used to prepare the signal for demodulation, but RF signal demodulation (to recover the original encoded information) is done by a demodulator.

(9) (B) NAK.

In an Automatic Repeat reQuest (ARQ) system, NAK (Negative Acknowledgment) is the signal the receiver sends to request the retransmission of a packet. ARQ is an error control method used in data communication to ensure reliable transmission. When a receiver detects an error in a received packet or fails to receive a packet within a specified time, it sends a NAK message back to the sender. Upon receiving the NAK, the sender retransmits the requested packet. This process continues until the receiver correctly receives the packet and sends an ACK (Acknowledgment) message to the sender. ACK is the signal sent by the receiver to confirm the successful reception of a packet. SYN (Synchronize) is a signal used to initiate a connection between two devices in a network. FIN (Finish) is a signal used to terminate a network connection.

(10) (B) Collector.

In an NPN transistor, the collector is usually connected to a positive voltage, while the emitter is connected to a negative or ground voltage. When a small current flows into the base, it allows a larger current to flow from the collector to the emitter, enabling the transistor to amplify or switch signals.

(11) (C) AND gate.

An AND gate is a digital logic gate that produces a high output only when all its inputs are high. In Boolean algebra, the AND function is represented by a dot (·) or by the absence of an operator. For example, the expression A AND B is written as $A \cdot B$ or simply AB. An OR gate produces a high output when at least one of its inputs is high. A NOR gate produces a low output when at least one of its inputs is high. A NAND gate produces a low output only when all its inputs are high.

(12) (D) Product detector.

In a single sideband (SSB) receiver, a product detector is used to demodulate the received signal by mixing it with a locally generated carrier. Single sideband modulation is a technique where either the upper or lower sideband of an AM signal is transmitted while the carrier and other sideband are suppressed. It must be mixed with a carrier signal at the original carrier frequency to demodulate an SSB signal. The product detector performs this mixing operation. It produces an output signal containing the original information signal and unwanted high-frequency components. These unwanted components are then filtered out from the demodulated information signal. A balanced modulator generates an SSB signal by suppressing the carrier and one of the sidebands. It is not used for demodulation. A phase detector is used in phase-locked loop (PLL) circuits to compare the phases of two signals and generate an error signal that can be used to adjust the frequency of one of the signals. A frequency discriminator is a circuit that converts frequency variations into amplitude variations.

(13) (B) Germanium.

Germanium diodes usually have a forward voltage drop of around 0.3 V, which matches the observed voltage drop in this scenario. A forward-biased diode is one where the positive terminal of the voltage source is connected to the anode (p-type), and the negative terminal is connected to the cathode (n-type). This allows current to flow easily through the diode. The forward voltage drop is the voltage difference across the diode when it conducts current in the forward direction. Silicon diodes have a higher forward voltage drop, usually around 0.7 V. Although there can be some variation, a voltage drop of 0.3 V is too low for a standard silicon diode. Zener diodes are designed to operate in reverse bias (the opposite of forward bias) and are not usually used for their forward voltage drop characteristics. Schottky diodes have a lower forward voltage drop than silicon diodes, but a 0.3 V drop is more characteristic of a germanium diode.

(14) (C) They are unsuitable for high-frequency applications.

Ferrite cores, especially those made of materials like nickel-zinc or manganese-zinc ferrites, are well-suited for high-frequency applications. Ferrite is a ceramic material with magnetic properties that make it ideal for inductor cores. Toroidal inductors are coils of wire wound around a donut-shaped core. The toroidal shape and ferrite core material provide several advantages for high-frequency use. Ferrite cores exhibit lower core losses (energy dissipated as heat) than iron cores at high frequencies, which makes them more efficient. They have high permeability (the ability of a material to support the formation of a magnetic field), which allows compact inductors with higher inductance values. They also maintain their properties well at high frequencies, which makes them suitable for various applications.

(15) (B) Resistance to current flow caused by inductors and capacitors.

Reactance is the opposition to current flow in an AC circuit that is specifically caused by inductors and capacitors. Unlike resistance, which remains constant regardless of frequency, reactance varies with frequency. Inductive reactance increases with higher frequencies, while capacitive reactance decreases with higher frequencies.

(16) (B) 16.2 watts.

Power dissipation in a resistor is the rate at which electrical energy is converted into heat. It is calculated using the formula: $P = E^2 \div R$, where P is the power dissipated in watts (W), E is the voltage across the resistor in volts, and R is the resistance of the resistor in ohms (Ω). In this case, the voltage across the resistor is 9 V, and the resistance is 5 Ω.

> Substitute the values.
> $P = (9)^2 \div 5$
> $P = 81^2 \div 5$
> $P = 16.2$ W

(17) (A) 120, mutual inductance.

The fundamental principle behind the operation of a transformer is mutual inductance. When alternating current flows through the primary winding, it creates a changing magnetic flux. This magnetic flux links with the secondary winding and induces a voltage across it. The ratio of voltages in a transformer is directly proportional to the ratio of the number of turns in the primary and secondary windings. This is mathematically represented as Vp ÷ Vs = Np ÷ Ns, where Vp = primary voltage, Vs = secondary voltage, Np = number of turns in the primary winding, and Ns = number of turns in the secondary winding. In this case, Vp = 120 V, Vs = 24 V, Np = 600 turns, while Ns is unknown.

> Make Ns the subject of the equation and substitute the known values.
> 120 V ÷ 24 V = 600 turns ÷ Ns
> Ns = 600 turns ÷ (120 V ÷ 24 V)
> Ns = 600 turns ÷ 5 V
> Ns = 120 turns

Electromagnetic induction is the principle behind electricity generation in a changing magnetic field. Lenz's law describes the direction of the induced current in a coil. They do not specifically address the interaction between two coils in a transformer. Ferromagnetic induction is not a standard term in the context of transformers. Ferromagnetic materials are used in the core of a transformer to enhance the magnetic flux, but they do not directly cause the voltage to appear in the secondary winding.

(18) (C) Signal distortion.

Noise reduction features in receivers analyze the incoming signal and attempt to identify and remove noise components. However, these algorithms are imperfect and can sometimes mistakenly remove parts of the desired signal and the noise. Overusing the noise reduction feature can remove too much of the original signal and produce distorted or unnatural-sounding audio. Frequency instability is usually caused by issues with the transmitter or receiver's oscillator, not the noise reduction feature. Noise reduction algorithms slightly reduce the receiver's sensitivity as they filter out some of the incoming signal with accompanying noise. Noise reduction often requires additional processing, which could slightly increase power consumption rather than decrease it.

(19) (B) Oscilloscope and digital multimeter.

An oscilloscope is the most suitable tool to examine the detailed shape of an RF signal, as it provides a real-time visual representation of the waveform. A digital multimeter (DMM) is designed to measure electrical quantities such as voltage, current, and resistance. In this case, the DMM will be used to measure the peak voltage of the RF signal, which is important to assess its power and potential for causing interference. A signal generator produces a known test signal, while a spectrum analyzer displays the frequency components of a signal. An antenna analyzer and a directional wattmeter are used to evaluate antenna performance and measure forward and reflected power. A field strength meter measures the intensity of an electromagnetic field at a specific location, while a dummy load is used to absorb RF power during transmitter testing. None of these tools is designed to examine an RF signal's detailed shape or peak voltage.

(20) (C) A humming sound in the audio.

A ground loop is a common problem in audio systems where multiple paths to the ground create a loop of electrical current. This loop can pick up electromagnetic interference, most often from the AC power lines, which manifests as a 60 Hz (or 50 Hz in some regions) humming sound in the audio output. SWR (Standing Wave Ratio) measures how well an antenna matches a transmission line. A high SWR can indicate a problem with the antenna system, but it is not directly related to ground loops in the audio system. Various factors, such as loose connections, atmospheric conditions, or electrical interference, can cause intermittent static noise. However, it is not a specific symptom of a ground loop. Ground loops mainly affect the receiving side of a radio system by introducing noise, not the transmit side.

(21) (B) The S meter will display S9+20 dB for the second signal.

The S-meter on a radio receiver is a relative indicator of signal strength, measured in S-units. The S-meter is not a precise measurement tool but rather a relative indicator that provides a general idea of how strong a signal is. The S-meter scale is logarithmic, not linear. This means that each S-unit represents a 6 dB (decibel) change in signal strength. A 6 dB increase in signal strength corresponds to doubling the received power, while a 6 dB decrease means the power is halved. The scale usually ranges from S1 (weakest) to S9 (strongest). Signals stronger than S9 are often indicated as S9+10 dB, S9+20 dB, and so on. A change of 20 dB corresponds to a 100-fold increase in power [10^(20 ÷ 10) = 100]. Therefore, if the first signal measures S9, a signal 100 times more powerful will be 20 dB stronger and produce an S-meter reading of S9+20 dB.

(22) (C) Blocking diode.

A blocking diode (check valve) is a one-way electrical device that only allows current to flow in one direction. In a solar power system, it is placed between the solar panel and the battery. In this configuration, the blocking diode prevents the battery from discharging back through the solar panel at night or during periods of low sunlight. An inverter converts direct current (DC) electricity from solar panels or batteries into alternating current (AC) electricity, which is used in most homes and businesses. It does not prevent battery discharge. A charge controller regulates the voltage and current from the solar panel to the battery to prevent overcharging and damage. It does not directly block reverse current flow. A voltage regulator maintains a constant voltage output from a power source, like a battery. It does not prevent the battery from discharging back through the solar panel.

(23) (C) Scatter propagation can bring radio signals to skip zones.

The skip zone is an area where radio signals cannot be directly received due to the nature of wave propagation. It exists between the range of the ground wave (which travels along the Earth's surface) and the sky wave (which is refracted back to Earth by the ionosphere). The skip zone has no direct relationship with the F2 layer. Scatter propagation allows signals to reach the skip zone by scattering off irregularities in the ionosphere or troposphere. Although these signals are usually weaker and more distorted than direct signals, they can still provide some communication within the skip zone.

(24) (B) As close to the MUF as possible.

To achieve the longest possible skip distance, operators should operate as close to the MUF as possible without exceeding it. The MUF represents the highest frequency that can be reliably refracted back to Earth by the ionosphere for a single hop. Operating just below the MUF allows the signal to be refracted at a shallower angle to increase the skip distance. Exceeding the MUF will cause the signal to penetrate the ionosphere and be lost in space. The critical frequency is the highest frequency that can be refracted vertically back to Earth. Operating just above the critical frequency will result in a short skip distance. Propagating midway between the LUF and MUF might work for some distances but is not the optimal strategy to maximize skip distance. The Lowest Usable Frequency (LUF) is the minimum frequency needed for skywave propagation. Operating near the LUF will likely cause excessive D-layer absorption and reduce the skip distance.

(25) (A) It is the highest layer in the ionosphere.

The F2 layer is the highest in the ionosphere and supports the longest skip distances for HF radio waves due to its altitude and ionization characteristics. The F2 layer is located between 200 and 400 km above the Earth's surface. This higher altitude means radio waves can travel farther before being refracted back to Earth and have longer skip distances. Although the F2 layer is the most ionized (meaning it has the highest density of free electrons), this is not the main reason for its ability to enable long skip distances. The ionization level is vital for refraction, but the altitude is more directly related to determining the skip distance.

(26) (C) Winlink.

Winlink is designed for amateur radio operators to send and receive emails when internet access is limited or unavailable. It achieves this by using radio frequencies to connect to a network of volunteer-operated gateway stations that relay messages to and from the internet. Common Message Server (CMS) is a component within the Winlink system that stores and manages messages. It is not a standalone system for email access. PACTOR is a digital protocol used for data transfer, not an email system. Gateway refers to a station within the Winlink network that has an internet connection and acts as a bridge between radio users and the internet. Although essential for Winlink, it is not the system itself.

(27) (A) CQ CQ CQ DE [Callsign] K.

This is the standard calling sequence used in amateur radio to initiate a contact. CQ is a general call to any station, DE means from, and [Callsign] is the operator's unique identifier. K (or AR in some regions) is a prosign that indicates the end of a transmission and an invitation for others to respond. Using any format other than the generally accepted standard can confuse other operators. QRZ? means "Who is calling me?" This is used to respond to a CQ call, not when initiating one.

(28) (D) The operator reduces their transmitter power to minimize potential interference with the distress signal.

When a distress signal is heard on an amateur radio frequency, the immediate priority is to assist the station in distress and ensure their safety. Reducing the transmitter power is the least helpful action in the given scenario. Although seemingly helpful, it could hinder communication with the distressed station. In this scenario, the most helpful action is for the operator to cease the ongoing transmission immediately. This allows the distressed station to communicate with them without interference. Pay close attention to the distress call to gather information about the nature of the emergency and the location of the distressed station. Let the distressed station know that their call has been heard and that help is on the way. Then, contact the appropriate authorities like the Coast Guard, local law enforcement, or other relevant emergency services as quickly as possible.

(29) (D) QRL.

Q signals are three-letter codes that amateur radio uses for quick and standardized communication. QRL? is a specific Q signal that translates to "Is this frequency in use?" It is used to inquire whether a frequency is clear before transmitting to ensure an ongoing conversation (QSO) between other stations is not interrupted. QRN signal refers to natural atmospheric noise, which can make it difficult to hear signals clearly. QSL is used to acknowledge receipt of a transmission or to confirm that a contact has been made. QRS means "Shall I send more slowly?" or "Send more slowly." It is used when the receiving station has trouble copying the code at the current speed.

(30) (A) Requires more complex equipment.

Single sideband (SSB), whether USB or LSB, requires more complex transmission and reception equipment than amplitude modulation (AM). This is because SSB transmitters and receivers must filter out one of the sidebands and often suppress the carrier. These modifications require additional circuitry and components. Although SSB has advantages like narrower bandwidth, increased power efficiency, and improved signal-to-noise ratio, the complexity of the equipment is a trade-off.

(31) (A) Cease transmission immediately and investigate the cause of the interference.

If an amateur radio operator causes interference to another service, such as television, their first responsibility is to cease transmission immediately. This is to prevent further disruption and to investigate the cause. They can adjust the antenna or reduce power to help, but it will not address the immediate issue. Transmitting outside amateur bands is illegal and could worsen the problem.

(32) (A) Yes, with a Certificate of Successful Completion of Examination (CSCE).

Upon passing the Amateur Extra exam, candidates are awarded a Certificate of Successful Completion of Examination (CSCE). This certificate grants hams temporary operating privileges of the new license class while the FCC processes their application and issues the physical license. The 15-meter band (21.000-21.450 MHz) is available for use by Amateur Extra Class licensees. There are no specific power restrictions tied to operating with a CSCE. Hams can operate at the maximum power allowed for their station and license class.

(33) (C) Federal Communications Commission (FCC), peak envelope power (PEP).

The Federal Communications Commission (FCC) is the regulatory body that oversees and regulates all radio communications (including amateur radio) in the United States. The FCC sets the rules and regulations (including power limit) for amateur radio operations. Peak Envelope Power (PEP) is the standard unit of measurement for these limits. PEP (Peak Envelope Power) is the maximum power of a radio signal during a single radio frequency (RF) cycle. It measures the highest power level the signal reaches at any given moment.

(34) (C) The regulations must be minimal and achieve a valid local objective.

Local authorities can regulate antenna installations, but these regulations must be minimal and serve a valid local purpose, such as preserving aesthetics or ensuring safety. They cannot be overly restrictive or prohibit amateur radio operations altogether. Local or state governments do not have the authority to sanction ham operators for exceeding 200 feet in antenna height. The FAA is the federal agency that regulates airspace and ensures the safety of air navigation. Local authorities cannot regulate antenna installations arbitrarily.

(35) (C) It is shared with other services and thus channelized.

The 60-meter band is unique as it is shared with other radio services, such as government and military operations. It is divided into specific channels that amateur radio operators must use to prevent interference. This channelization contrasts with other bands, where operators have more freedom to choose their frequencies. The 60-meter band also has a maximum power output of 100 watts ERP (Effective Radiated Power), which is lower than many other bands.

Test 4: Questions

(1) What is admittance in an AC circuit?

(A) The reciprocal of impedance, representing how easily current flows.

(B) The opposition to voltage in a DC circuit.

(C) A measure of power in an AC circuit.

(D) The resistance provided by inductors in an AC circuit.

(2) Which statement most accurately describes the allowed frequencies for automatically controlled beacons?

(A) They are allowed anywhere on the HF bands if the station licensee provides prior notification to the FCC.

(B) They are only permitted within designated frequency ranges on the 10-meter band.

(C) They are restricted to the 20-meter band for propagation studies.

(D) They are allowed on frequencies 26.20 MHz to 26.30 MHz.

(3) According to the FCC, what is the requirement for operators when using an antenna with a higher gain than a dipole on 60 meters?

(A) Inform nearby stations of the antenna strength.

(B) Keep a record of the frequency at which they operate.

(C) Keep a record of the antenna's gain.

(D) Regularly inspect the antenna for wear and tear.

(4) Jake Burrow just passed his General Class license examination. His upgrade has not been effected, but he is now eager to transmit on General Class frequencies. How must Jake identify himself on air?

(A) His current callsign, as his Technician license is still valid.

(B) His current callsign with "/AG" appended at the end.

(C) Any callsign he chooses, as the FCC has not assigned one yet.

(D) He cannot transmit on General Class frequencies until he receives his new license.

(5) In the 2.4 GHz band, what type of user designation do amateur radio operators have?

(A) Primary users.

(B) Secondary users.

(C) Tertiary users.

(D) Unlicensed users.

(6) When operating on the 20-meter band, an operator hears a station calling "CQ DX." What does this indicate?

(A) The station is looking for local contacts within their country.

(B) The station is only interested in contacting stations from Europe.

(C) The station is looking for contacts outside their country.

(D) The station is participating in a contest and needs to make as many contacts as possible.

(7) An operator tunes their radio for a CW (Morse code) transmission and notices another station actively communicating nearby. To avoid interfering with their signal, which frequency separation should they aim for when selecting a transmitting frequency?

(A) 50 Hz.

(B) 300 Hz.

(C) 1 kHz.

(D) 5 kHz.

(8) During a CW QSO, atmospheric noise makes it difficult to copy the other station. Which Q signal best conveys this situation?

(A) QRM.

(B) QRN.

(C) QSB.

(D) QRS.

(9) An operator's callsign is AB2VNU, and they are trying to contact their friend Vance over a noisy channel. How will they pronounce their callsign using the NATO phonetic alphabet?

(A) Able Baker Two Victor November Uniform.

(B) Alpha Bravo Two Victor November Uniform.

(C) Alpha Bravo Two Vance November Uniform.

(D) Able Baker Two Victory November Uniform.

(10) Which digital mode is most commonly found in the frequency range of 14.074 MHz to 14.077 MHz?

(A) PSK31.

(B) JT65.

(C) FT8.

(D) Olivia.

(11) What is the maximum angle at which a radio wave can be transmitted and still be returned to Earth by the ionosphere?

(A) Angle of incidence.

(B) Brewster's angle.

(C) Angle of refraction.

(D) Critical angle.

(12) Radio waves with frequencies above the MUF are most likely to do what?

(A) Be refracted back to Earth by the F2 layer.

(B) Pass through the ionosphere into space.

(C) Be absorbed by the D layer.

(D) Undergo multiple hops between the Earth and the ionosphere.

(13) Which statement regarding geomagnetic storms is true?

(A) Geomagnetic storms enhance HF propagation at higher latitudes.

(B) The Earth's magnetic field is strengthened during a geomagnetic storm.

(C) Geomagnetic storms only affect low-frequency bands.

(D) Geomagnetic storms can disrupt HF propagation at higher latitudes.

(14) In a solar panel, how are multiple photovoltaic cells interconnected?

(A) Parallel connections to increase the total current output.

(B) Series connections to increase the total voltage output.

(C) Both series and parallel connections.

(D) A complex network that maximizes power output under varying conditions.

(15) An operator operates a ham radio station on the 20-meter band and wants to transmit a single sideband (SSB) signal with a lower sideband (LSB) bandwidth of 5 kHz. The radio display shows a carrier frequency of 14.200 MHz. What range of frequencies should they maintain to avoid interfering with other transmissions?

(A) 14.195 MHz to 14.200 MHz.

(B) 14.200 MHz to 14.205 MHz.

(C) 14.190 MHz to 14.205 MHz.

(D) 14.195 MHz to 14.205 MHz.

(16) What is the main purpose of bonding metal equipment enclosures in an amateur radio station?

(A) To improve antenna efficiency.

(B) To reduce harmonic emissions.

(C) To reduce RF voltage differences between enclosures.

(D) To prevent power supply overload.

(17) An operator needs to verify the linearity of a power amplifier in a transmitter. Which test will be the most suitable for this purpose?

(A) Two-tone test.

(B) Keying waveform analysis.

(C) RF envelope pattern measurement.

(D) SWR measurement.

(18) What is the primary function of the TX Delay setting on a transceiver when using an external amplifier?

(A) To prevent SWR damage.

(B) To reduce harmonic distortion.

(C) To synchronize the amplifier with the transceiver.

(D) To allow the amplifier to switch to transmit mode.

(19) Four resistors with values of 5 ohms, 10 ohms, 15 ohms, and 20 ohms are first connected in parallel, then reconnected in series. What is the total equivalent resistance when connected in parallel, and what is it when in series?

(A) Parallel: 2.4 ohms, Series: 50 ohms.

(B) Parallel: 50 ohms, Series: 2.4 ohms.

(C) Parallel: 12.5 ohms, Series: 25 ohms.

(D) Parallel: 25 ohms, Series: 12.5 ohms.

(20) A circuit component with a resistance of 2.2 kilohms has a current of 10 milliamps flowing through it. How much power is dissipated in the component?

(A) 22 milliwatts.

(B) 220 milliwatts.

(C) 2.2 watts.

(D) 22 watts.

(21) What statement about electrical properties in AC circuits is untrue?

(A) Impedance is the total opposition to the flow of alternating current.

(B) Admittance is the reciprocal of impedance.

(C) Impedance is calculated as the ratio of voltage to current.

(D) Admittance is measured in ohms.

(22) For a high-frequency radio transmitter operating at 5 GHz, which connector type will provide the best waterproofing to attach the antenna cable with minimal signal loss?

(A) Type F.

(B) Type N.

(C) BNC.

(D) UHF.

(23) Which of the statements about wire-wound resistors is true?

(A) They are ideal for high-frequency circuits due to their low inductance.

(B) They are commonly used in low-power applications due to their small size.

(C) Their erratic behavior can negatively affect circuits at high frequencies.

(D) They are known for their low tolerance values and are often used in precision circuits.

(24) What best describes insertion loss?

(A) The decrease in signal amplitude caused by the filter within its passband.

(B) The frequency at which the filter starts to attenuate the signal.

(C) The ripple or variation in attenuation within the passband.

(D) The filter's maximum attenuation to signals outside its passband.

(25) How many distinct states can a 4-bit binary counter represent?

(A) Four.

(B) Eight.

(C) Sixteen.

(D) Thirty-two.

(26) What is the circuit element shown?

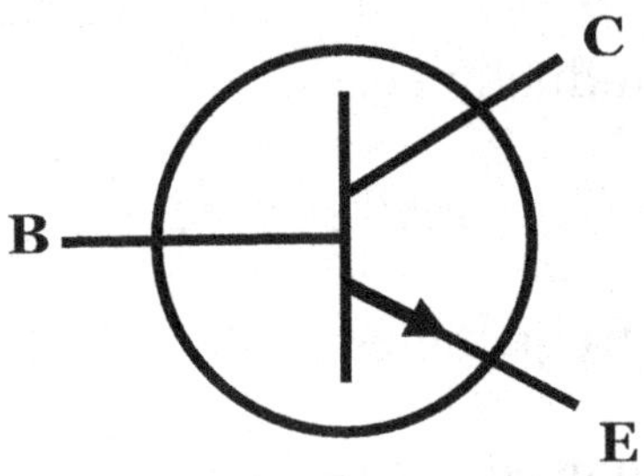

(A) NPN junction transistor.

(B) Solid core transformer.

(C) Field effect transistor.

(D) Zener diode.

(27) Which statement about upper-case letters in PSK31 is true?

(A) They increase transmission speed.

(B) They slow down transmission.

(C) They do not affect transmission speed.

(D) Uppercase fonts improve error correction.

(28) What is the function of a multiplier in a transmitter?

(A) To combine two or more signals of different frequencies.

(B) To filter out unwanted harmonics from a signal.

(C) To increase the frequency of a signal by generating integer multiples of the input frequency.

(D) To amplify the power of a signal.

(29) Which statement accurately describes Amplitude Modulation (AM)?

(A) It varies the phase angle of the carrier wave to convey information.

(B) It varies the instantaneous frequency of the carrier wave to convey information.

(C) It varies the amplitude (strength) of the carrier wave to convey information.

(D) It requires oscillators and mixers connected in series.

(30) Which of the statements below is incorrect?

(A) A log-periodic antenna can operate over a wide range of frequencies.

(B) A screwdriver mobile antenna adjusts its impedance by changing base loading inductance.

(C) A Beverage antenna is mainly used for directional reception of signals in the medium-frequency and lower high-frequency bands.

(D) The spacing of elements in a log-periodic antenna is uniform along its length.

(31) An amateur radio operator experiences interference from a station located behind their Yagi antenna. Which antenna specification will be most relevant in determining how effectively the antenna can reject this unwanted signal?

(A) Gain in dBi.

(B) Beamwidth.

(C) Front-to-back ratio.

(D) SWR bandwidth.

(32) What happens to the azimuthal radiation pattern of a horizontal dipole HF antenna when it is less than half a wavelength high?

(A) The pattern becomes highly directional.

(B) The pattern becomes almost omnidirectional.

(C) The pattern develops deep nulls.

(D) The pattern shifts to a higher elevation angle.

(33) What Standing Wave Ratio results from connecting a 90-ohm feed line to a 30-ohm resistive load?

(A) 3:1.

(B) 1:3.

(C) 4:1.

(D) 1:4.

(34) What is the most important safety consideration when operating an emergency generator?

(A) Ensure the generator is fueled with the correct type of gasoline.

(B) Provide adequate ventilation to prevent carbon monoxide buildup.

(C) Ground the generator to a metal water pipe.

(D) Place the generator on a level surface to prevent it from tipping over.

(35) Which statement is not correct?

(A) RF exposure evaluation must follow FCC OET Bulletin 65.

(B) Lower duty cycle allows higher power levels.

(C) Calibrated equipment can be used to determine FCC RF exposure compliance.

(D) Time averaging is the total RF exposure time.

Test 4: Answers and Explanations

(1) (A) The reciprocal of impedance, representing how easily current flows.

Admittance is the measure of how easily current flows in an AC circuit and is the reciprocal of impedance. Measured in siemens, admittance accounts for both the resistance and reactance in the circuit.

(2) (B) They are only permitted within designated frequency ranges on the 10-meter band.

Amateur radio operators can establish beacon stations on the 10-meter band (28.2-28.3 MHz) without special authorization. Transmissions must be automatically controlled and include only their call sign, location, and the designated power limit. Beacon stations are not allowed anywhere on the HF bands. They are restricted to specific frequency ranges to avoid interference with other radio services.

(3) (C) Keep a record of the antenna's gain.

The FCC requires that if an antenna other than a half-wave dipole is used at 60 meters, the user must keep a record of the antenna's gain. This requirement stems from the fact that the 60-meter band is a secondary allocation for amateur radio operators, which means they must not cause harmful interference to primary users. The FCC limits the effective radiated power (ERP) on 60 meters to 100 watts PEP relative to a half-wave dipole. Since a high-gain antenna can increase the ERP beyond this limit even with a lower power output, operators using such antennas must be aware of their actual ERP and ensure it does not exceed the allowed limit. Keeping a record of the antenna's gain is essential to calculate the ERP and ensure compliance with FCC regulations.

(4) (B) His current callsign with "/AG" appended at the end.

When a US amateur radio operator passes a higher-class license exam (like the General exam in this case) but has not yet received their new license from the FCC, they are granted temporary operating privileges on the new bands and modes. This is confirmed by the Certificate of Successful Completion of Examination (CSCE) they receive after passing. To indicate they are operating with these upgraded privileges, they must append "/AG" (Advanced General) to their existing callsign when transmitting on those new frequencies or modes. For example, if Jake's current callsign were KI5XYZ, he would identify himself on air as KI5XYZ/AG. This lets other operators know he is legally operating on that frequency, even though his license has not been formally upgraded yet.

(5) (B) Secondary users.

In the 2.4 GHz band, which is also used by various wireless devices like Wi-Fi and Bluetooth, amateur radio operators are considered secondary users. This means they are allowed to use the band, but they must not cause harmful interference to primary users, who have priority access. Primary users in this band include certain industrial, scientific, and medical (ISM) applications. If interference occurs, secondary users must take steps to mitigate it or cease operation.

(6) (C) The station is looking for contacts outside their country.

The term "DX" in amateur radio is short for distance and refers to long-distance communication, especially with stations in other countries or continents. A station calling "CQ DX" is essentially saying, "Calling any station, anywhere in the world, for a long-distance contact." This is a common way for amateur radio operators to seek out communication with stations far away from their location. This practice helps hams expand their reach and make contacts with people from different cultures and backgrounds.

(7) (B) 300 Hz.

In CW (Morse code) communication, signals are transmitted as a series of short and long tones (dots and dashes). These tones occupy a certain bandwidth on the radio frequency spectrum. A minimum frequency separation is necessary to avoid interference between adjacent CW signals. The recommended standard for minimum frequency separation between CW signals is between 150 Hz and 500 Hz. This range allows clear reception of both signals without significant overlap or distortion. 300 Hz falls comfortably within this recommended range, which makes it the most suitable minimum separation in this scenario. However, experienced operators may use even wider spacing, especially when working with weaker signals or in crowded band conditions where interference is more likely. The optimal frequency separation depends on various factors, such as the operator's skill level, the quality of the equipment, and the specific conditions on the band.

(8) (B) QRN.

Atmospheric noise, or static, is natural interference caused by lightning and other atmospheric phenomena. It manifests as random crackling or hissing sounds in the receiver. QRN (N for natural) indicates the presence of this type of noise. QRM (M for man-made) indicates interference from other stations. QSB (fading signals) describes fading signals, where the strength of the received signal fluctuates due to atmospheric conditions.

(9) (B) Alpha Bravo Two Victor November Uniform.

The NATO phonetic alphabet, also known as the International Radiotelephony Spelling Alphabet, is a set of standardized code words used to represent letters of the alphabet. Its purpose is to ensure unambiguous communication over voice channels, especially in noisy or unclear conditions. NATO alphabets are used to communicate with people who may have different native languages and accents. In amateur radio, the NATO phonetic alphabet is commonly used to spell out call signs, names, and other important information to ensure accurate reception and understanding. Using the NATO phonetic alphabet helps to avoid confusion and misinterpretation, especially with letters that sound similar, such as /b/ and /v/ or /m/ and /n/. In this specific case, the call sign AB2VNU will be spelled out as:

- **A**: Alpha.
- **B**: Bravo.
- **2**: Two.
- **V**: Victor.
- **N**: November.
- **U**: Uniform.

(10) (C) FT8.

FT8 is a weak-signal digital mode known for its ability to decode signals with extremely low signal-to-noise ratios. It is mainly designed to make quick contacts and exchange basic information, such as call signs, signal reports, and grid locators. Due to its efficiency and robustness, FT8 has become popular among amateur radio operators. It is often used on the 20-meter band (14 MHz), commonly found in the frequency range of 14.074 MHz to 14.077 MHz. PSK31 is another digital mode used for keyboard-to-keyboard communication. Although it operates on HF frequencies, it is usually found on different parts of the bands than FT8. JT65 is a weak-signal mode similar to FT8, but it is generally used for EME (Earth-Moon-Earth) communication and other long-distance contacts under challenging conditions. Olivia is a digital mode designed for higher data rates and is often used to transmit images and other files over radio.

(11) (D) Critical angle.

The critical angle is the maximum angle at which a radio wave can be transmitted from the Earth's surface and still be refracted back down by the ionosphere. Suppose the angle of incidence (the angle between the transmitted wave and a line perpendicular to the Earth's surface) exceeds the critical angle. In that case, the wave will not be bent enough to return and pass through the ionosphere. The angle of incidence is the angle at which the wave strikes the ionosphere, but not the maximum angle for refraction. Brewster's angle relates to the polarization of light waves at a dielectric interface and does not apply to radio wave propagation in the ionosphere. The angle of refraction is the angle at which the wave bends within the ionosphere, but not the maximum angle for returning to Earth.

(12) (B) Pass through the ionosphere into space.

Radio waves with frequencies above the MUF are not refracted back to Earth by the ionosphere. Instead, they pass through the ionosphere and continue out into space. The ionosphere's ability to refract radio waves decreases as the frequency increases. The MUF represents the frequency at which the refraction is sufficient to return the signal to Earth. Above this frequency, the refraction angle is too small, and the signal escapes into space.

(13) (D) Geomagnetic storms can disrupt HF propagation at higher latitudes.

Geomagnetic storms are disturbances in the Earth's magnetic field caused by solar events like coronal mass ejections (CMEs). These storms can disrupt HF (high frequency) radio propagation, especially at higher latitudes closer to the poles. The disturbances can cause fluctuations in the ionosphere. They cause scattering, absorption, and other effects that degrade or completely disrupt HF signals. Geomagnetic storms generally degrade HF propagation, not enhance it. They weaken the Earth's magnetic field and can affect a wide range of frequencies, not just low-frequency bands.

(14) (C) Both series and parallel connections.

Solar panels consist of multiple individual photovoltaic (PV) cells, each of which generates a small amount of electrical current and voltage when exposed to sunlight. These cells are interconnected with series and parallel configurations to achieve the desired voltage and current output for a solar panel. When PV cells are connected in series, their voltages add up. This increases the overall voltage output of the solar panel. When PV cells are connected in parallel, their currents add up. This increases the overall current output of the solar panel. Solar panel manufacturers can tailor the output voltage and current to match the requirements of different applications by using a combination of series and parallel connections. For example, a panel designed to charge a 12-volt battery might have several strings of cells connected in series to achieve the necessary voltage, with each string consisting of cells connected in parallel to increase the current output.

(15) (A) 14.195 MHz to 14.200 MHz.

In single sideband (SSB) transmission, either the upper or lower sideband of the modulated signal is transmitted, while the carrier and the other sideband are suppressed. This produces a more power-efficient and narrower bandwidth signal than traditional amplitude modulation (AM). In LSB, the lower sideband is transmitted, and the upper sideband is suppressed. The carrier frequency represents the upper edge of the transmitted signal's spectrum. To determine the signal's frequency range, subtract the bandwidth from the carrier frequency. In this question, the operator transmits an LSB signal with a bandwidth of 5 kHz and a carrier frequency of 14.200 MHz. Therefore, the frequency range of the signal will be from 14.195 MHz (14.200 MHz-5 kHz) to 14.200 MHz. That is, they transmit within a range of 5 kHz below the carrier frequency displayed by their radio dial.

(16) (C) To reduce RF voltage differences between enclosures.

Bonding electrically connects the metal chassis and enclosures of various pieces of equipment in an amateur radio station. The main purpose is to create a common electrical potential throughout the station, which minimizes the potential differences in radio frequency (RF) voltage between various equipment enclosures. This is essential for RF safety, interference reduction, and lightning protection. RF voltages can build up on ungrounded metal surfaces and pose a risk of electric shock to the operator. Bonding reduces this risk by ensuring that all exposed metal surfaces are at the same potential. RF currents flowing between equipment enclosures can cause interference with other electronic devices and even with the radio equipment itself. Bonding minimizes these RF currents and facilitates a cleaner signal and reduced noise. It helps to equalize the electrical potential across all bonded equipment during a nearby lightning strike.

(17) (A) Two-tone test.

A two-tone test is a specialized technique to assess the linearity of a power amplifier. It involves injecting two closely spaced audio tones into the amplifier's input and observing the output spectrum. A linear amplifier will produce only the two original tones in the output, while a nonlinear amplifier will generate additional intermodulation products (undesired signals at different frequencies). The presence and strength of these intermodulation products can be used to quantify the amplifier's linearity. Keying waveform analysis examines the shape of the on-off keying waveform used for Morse code transmission. Although important for CW operation, it does not directly assess amplifier linearity. The RF envelope pattern measurement technique visualizes the overall shape of the RF signal envelope, which can reveal some types of distortion. However, it is not as sensitive or specific as a two-tone test for measuring amplifier linearity. SWR (Standing Wave Ratio) measures how well an antenna matches a transmission line. Although SWR can indirectly impact amplifier performance, it is not a direct measure of linearity.

(18) (D) To allow the amplifier to switch to transmit mode.

When using an external amplifier with a transceiver, there is often a slight delay between when the transceiver initiates the transmit sequence and when the amplifier is fully ready to amplify the signal. The TX Delay setting on the transceiver introduces a short delay (usually a few milliseconds) between the transceiver's transmit command and the actual RF output. This allows the amplifier to switch from receive to transmit mode before applying the RF power. It prevents the initial portion of the transmission from being clipped or distorted. SWR (Standing Wave Ratio) protection is a separate function, usually done by the amplifier itself or an external SWR meter. Harmonic distortion is mainly a concern in the amplifier's design, not the TX Delay setting on the transceiver. TX Delay is not for precise synchronization with the transceiver's internal timing.

(19) (A) Parallel: 2.4 ohms, Series: 50 ohms.

In a parallel circuit, the total equivalent resistance is calculated using the following formula: $1/R_{total} = 1/R1 + 1/R2 + 1/R3 + 1/R4$.

Substitute the values.
$1/R_{total} = 1/5 + 1/10 + 1/15 + 1/20$
$1/R_{total} = (12 + 6 + 4 + 3) \div 60$
$1/R_{total} = 25 \div 60$
$1/R_{total} = 60 \div 25$
$R_{total} \approx 2.4$ ohms

In a series circuit, the total equivalent resistance is simply the sum of the individual resistances: $R_{total} = R1 + R2 + R3 + R4$.

Substitute the values.
$R_{total} = 5 + 10 + 15 + 20$
$R_{total} = 50$ ohms.

(20) (B) 220 milliwatts.

Power dissipation in an electrical component is the rate at which electrical energy is converted into other forms, such as heat, light, or mechanical work. In a resistor, the power dissipated is converted into heat. It can be calculated using the formula: $P = I^2 \times R$, where P is the power dissipated in watts (W), I is the current flowing through the resistor in amperes (A), and R is the resistance of the resistor in ohms (Ω). In this case, the resistance is 2.2 kilohms, which needs to be converted to ohms: 2.2 kilohms = 2200 ohms. The current is 10 milliamps, which needs to be converted to amps: 10 milliamps = 0.01 amps.

Substitute these values.
$P = (0.01\ A)^2 \times 2200\ \Omega = 0.22\ W = 220\ mW$

(21) (D) Admittance is measured in ohms.

Admittance (Y) measures how easily alternating current (AC) flows in a circuit. It is the reciprocal of impedance (Z). Impedance is the total opposition to the flow of alternating current. Impedance is measured in ohms (Ω), while admittance is measured in Siemens (S). The impedance formula for AC circuits, $Z = V/I$, is an extension of Ohm's Law, which is a fundamental principle in electrical engineering. Ohm's Law states that the current (I) flowing through a conductor is directly proportional to the voltage (V) applied across it and inversely proportional to the resistance (R) of the conductor. In DC circuits, $V = I \times R$, where V is the voltage in volts (V), I is the current in amperes (A), and R is the resistance in ohms (Ω). This can be mathematically rearranged as $R = V \div I$. In AC circuits, impedance (Z) replaces resistance (R), and the formula becomes $Z = V \div I$, where Z is the impedance in ohms (Ω), V is the voltage in volts (V), and I is the current in amperes (A).

(22) (B) Type N.

Type N connectors are the best choice for attaching the antenna cable to a high-frequency radio transmitter operating at 5 GHz when considering waterproofing and minimal signal loss. Type N connectors are coaxial connectors designed for use in high-frequency applications. They provide excellent performance up to 11 GHz. They have a threaded coupling mechanism, which provides secure and reliable connections. In addition, they have a characteristic impedance of 50 ohms, commonly used in RF applications. Type F connectors are commonly used for cable television and satellite connections. BNC connectors are widely used for lower-frequency applications and may not be weather-resistant. UHF connectors are older and less commonly used for modern high-frequency applications. They do not have the same level of performance and waterproofing as Type N connectors.

(23) (C) Their erratic behavior can negatively affect circuits at high frequencies.

Wire-wound resistors are constructed by winding a resistive wire around a core, which creates an inherent inductance. Inductance is the tendency of an electrical conductor to oppose changes in the current flowing through it. Although this inductance is negligible at low frequencies, it becomes significant at high frequencies. The resulting inductive reactance can cause the resistor to behave erratically and deviate from its intended resistance value. Thus, standard wire-wound resistors are generally unsuitable for high-frequency circuits. Wire-wound resistors come in various sizes, and their size is not the primary factor that determines their suitability for low-power applications. They are often used in high-power applications where their ability to dissipate heat is advantageous. Wire-wound resistors are available with a range of tolerances, with some having low and others having high tolerance ratings.

(24) (A) The decrease in signal amplitude caused by the filter within its passband.

Insertion loss refers to the reduction in signal amplitude caused by a filter within its passband (intended frequency range). When a signal passes through a filter, some of its energy is lost due to the resistance and reactance of the filter components. This loss in signal power manifests as a decrease in amplitude. Insertion loss is usually expressed in decibels (dB) and is an important parameter to consider when designing or selecting filters. The frequency at which the filter starts to attenuate the signal is called the filter's cutoff frequency. The ripple or variation in attenuation within the passband is the passband ripple, which measures the filter's flatness in the passband. The filter's maximum attenuation to signals outside its passband is the stopband attenuation. It measures how effectively the filter rejects signals outside of its passband.

(25) (C) Sixteen.

A 4-bit binary counter can represent sixteen distinct states. A binary counter is a digital circuit that counts in binary numbers. Each bit in the counter can have two possible states (0 or 1), so a 4-bit counter has $2^4 = 16$ possible combinations of these states. The sixteen states a 4-bit binary counter can represent are 0000, 0001, 0010, 0011, 0100, 0101, 0110, 0111, 1000, 1001, 1010, 1011, 1100, 1101, 1110, and 1111.

(26) (A) NPN junction transistor.

The circuit symbol represents an NPN junction transistor. A transistor is a semiconductor device used to amplify or switch electronic signals and electrical power. Transistors are fundamental building blocks in modern electronics. Junction transistors are a type of transistor where the connections between the emitter and base and between the base and collector are formed by p-n junctions. An NPN transistor has three layers of semiconductor material: an N-type emitter, a P-type base, and an N-type collector. The arrow in the symbol points outward from the emitter. It indicates the direction of conventional current flow (positive to negative). NPN transistors are current-controlled devices, which means a small current flowing into the base controls a larger current flowing from the collector to the emitter. A transformer is a passive electrical device that transfers electrical energy between two or more circuits through mutual inductance. Its symbol usually consists of two coils with lines between them that represent the primary and secondary windings. Field effect transistors (FETs) are also three-terminal devices like transistors but are voltage-controlled. Their symbols are similar to that of a BJT, but the base terminal is replaced by a gate terminal, often represented by an arrow pointing inward for N-channel FETs and outward for P-channel FETs. A Zener diode operates in the reverse breakdown region to provide a stable reference voltage. Its symbol includes a line resembling a Z at the cathode end.

(27) (B) They slow down transmission.

In PSK31, a popular digital mode for text communication used in amateur radio, uppercase letters slow the transmission. PSK31 (Phase Shift Keying 31) is a type of digital modulation that uses varying phase shifts to represent digital data. It uses a variable-length symbol scheme called Varicode. In Varicode, more common characters are assigned shorter symbols, and less common characters are assigned longer symbols. Since uppercase letters are generally less frequent in text than lowercase letters, they are assigned longer symbols. This slows the overall transmission speed. Error correction in PSK31 is achieved through other mechanisms, such as checksums or forward error correction (FEC).

(28) (C) To increase the frequency of a signal by generating integer multiples of the input frequency.

A multiplier in a transmitter is a circuit designed to generate harmonics, which are integer multiples of the input frequency. For example, a frequency doubler takes an input signal at frequency f and produces an output signal at 2f. Similarly, a frequency tripler will produce an output signal at 3f, and so on. This frequency multiplication is achieved through nonlinear circuit elements, such as diodes or transistors, that introduce distortion to the input signal. This distortion creates additional frequency components at harmonics of the original frequency. A filter then selects the desired harmonic and removes the unwanted fundamental frequency and other harmonics. Other devices in a transmitter might be mistaken for a multiplier, but their functions are distinct. A mixer combines multiple signals of different frequencies. A filter, such as a bandpass or low-pass filter, removes unwanted harmonics from a signal. An amplifier also increases the power of a signal. Although some multipliers may introduce a slight gain, their main function is to increase the frequency, not amplify the signal's power.

(29) (C) It varies the amplitude (strength) of the carrier wave to convey information.

In Amplitude Modulation (AM), the information signal is encoded by varying the amplitude of the carrier wave. The amplitude of the carrier wave is directly proportional to the strength of the information signal at any given instant. Varying the phase angle of the carrier wave to convey information describes Phase Modulation (PM). Varying the instantaneous frequency of the carrier wave to convey information refers to Frequency Modulation (FM). Although oscillators and mixers are used in AM transmitters, they are not the defining characteristic of AM. The key aspect is the variation of the carrier wave's amplitude.

(30) (D) The spacing of elements in a log-periodic antenna is uniform along its length.

In a log-periodic antenna, the element spacing is not uniform along its length. Instead, it follows a logarithmic progression, with the distance between elements gradually increasing as they move away from the feed point. This unique design allows the log-periodic antenna to operate over a wide range of frequencies (making it a broadband antenna). Meanwhile, screwdriver mobile antennas use a variable inductor at their base to adjust their impedance for optimal matching across different frequencies. Beverage antennas, designed as long-wire antennas, are mainly employed for directional reception of signals in the medium frequency (MF) and lower high-frequency (HF) bands. This enables beverage antennas to be used for targeted reception from specific directions.

(31) (C) Front-to-back ratio.

The front-to-back ratio (F/B ratio) is the most relevant antenna specification for determining how effectively a Yagi antenna can reject unwanted signals from the rear. It quantifies the difference in signal strength (measured in decibels) between the antenna's forward radiation (main lobe) and its backward radiation (back lobe). A higher F/B ratio indicates better rejection of signals from the rear. In a Yagi antenna, the front-to-back ratio measures its directivity, which is the ability to focus the radiated energy in a particular direction. A higher directivity means the antenna is more focused on transmitting and receiving signals in the desired direction. It also means it does this while minimizing the reception of signals from other directions (including those from the rear).

(32) (B) The pattern becomes almost omnidirectional.

When a horizontal dipole HF antenna is less than half a wavelength high, its azimuthal radiation pattern becomes almost omnidirectional. The azimuthal radiation pattern describes how an antenna radiates energy in the horizontal plane (360 degrees around the antenna). A dipole antenna consists of two conductive elements, usually made of wire, each a quarter-wavelength long. At heights less than half a wavelength, the ground reflections interact with the direct radiation from the antenna in a way that fills in the nulls (directions of minimum radiation) in the pattern. This produces a more omnidirectional pattern, which means the antenna radiates energy more or less equally in all directions in the horizontal plane.

(33) (A) 3:1.

The Standing Wave Ratio (SWR) measures how well the impedance of an antenna system is matched to the impedance of the transmission line. It is calculated as the ratio of the maximum voltage to the minimum voltage along the transmission line. Equivalently, it is the ratio of the higher impedance to the lower impedance. For a 90-

ohm feed line and a 30-ohm antenna, the SWR will be (90 ÷ 30) 3:1. Since the feedline impedance (90 ohms) is higher, we divide the feedline impedance by the antenna impedance.

(34) (B) Provide adequate ventilation to prevent carbon monoxide buildup.

The most important safety consideration when operating an emergency generator is to provide adequate ventilation to prevent carbon monoxide buildup. Carbon monoxide (CO) is a colorless, odorless gas produced by the combustion of fuels like gasoline or propane. It is poisonous and can be fatal if inhaled in sufficient quantities. Generators should never be operated indoors or in enclosed spaces without proper ventilation.

(35) (D) Time averaging is the total RF exposure time.

Time averaging is not the total RF exposure time but rather a method used to assess RF exposure over a specified period. RF exposure limits are usually based on time-averaged values, which consider the varying intensity of RF fields over time. Time averaging measures the RF exposure over a certain period (for example, six minutes) and calculates the average power density. This average value is then compared to the exposure limits to determine compliance. This method is used because RF exposure can vary significantly depending on the transmitter's duty cycle and other factors. Total RF exposure time refers to the entire duration a person is exposed to RF fields, regardless of the intensity variations. For example, a ham radio operator might transmit at a high power level for one minute and then remain silent for five minutes. The total RF exposure time is six minutes, but the time-averaged exposure will be much lower than the peak exposure during transmission.

Test 5: Questions

(1) What is the requirement for a station transmitting time-averaged power above 0.5 milliwatts in a regulated band?

(A) It must use digital encoding.

(B) It must identify itself with a call sign.

(C) It must be coordinated through the FCC.

(D) It must avoid transmissions on the 10-meter band.

(2) Why is it essential to bond all ground rods together in a lightning protection system?

(A) To create a low-resistance path for lightning current to dissipate into the ground.

(B) To prevent the ground rods from rusting or corroding.

(C) To ensure the system is aesthetically pleasing.

(D) To protect the ground rods from theft or damage.

(3) What is the main consequence of high SWR on transmitted power?

(A) Increased power radiated from the antenna.

(B) Reduced power reaching the antenna for radiation.

(C) No change in the power reaching the antenna.

(D) Less power is reflected back to the power source.

(4) What is a key characteristic of a quarter-wave vertical antenna?

(A) It requires a ground plane or radial system for efficient operation.

(B) It is designed primarily for use on the VHF bands.

(C) It has a higher radiation angle than a dipole antenna.

(D) It requires no tuning for operation on multiple bands.

(5) Which part of a directional antenna produces the maximum radiated field strength?

(A) The antenna's tip.

(B) The most current-dense region.

(C) The director.

(D) The main lobe.

(6) What is the advantage of using an inverted-V dipole configuration?

(A) It increases the antenna's gain.

(B) It improves the antenna's front-to-back ratio.

(C) It conserves horizontal space.

(D) It eliminates the need for a balun.

(7) In an AM transmission, what will likely be caused by excessive audio input levels?

(A) Flat-topping.

(B) Carrier shift.

(C) Frequency drift.

(D) Spurious emissions.

(8) What is the frequency deviation of an 8.93 MHz oscillator used for reactance modulation in a 145.17 MHz FM phone transmitter with a maximum deviation of 3 kHz?

(A) 184.5 Hz.

(B) 3 kHz.

(C) 30 kHz.

(D) 48.4 MHz.

(9) How is overmodulation detected on a waterfall display?

(A) By the presence of horizontal lines.

(B) By the absence of any signal.

(C) By the signal appearing faint and weak.

(D) By one or more vertical lines on either side of the signal.

(10) Which circuit component is this?

(A) Zener diode.

(B) Tapped inductor.

(C) NPN junction transistor.

(D) Solid core transformer.

(11) Which digital circuit is designed to systematically move data bits through a series of storage elements under the control of a clock signal?

(A) Multiplexer.

(B) Demultiplexer.

(C) Shift register.

(D) Counter.

(12) Which of these statements is not correct?

(A) Digital signal processors can operate only one type of bandwidth.

(B) In I/Q modulation, the phase relationship between the in-phase (I) and quadrature (Q) signals is ninety degrees.

(C) Demodulator bandwidth can alter a receiver's sensitivity.

(D) Software-defined radios (SDR) can filter, modulate, demodulate, and detect signals.

(13) In a tetrode vacuum tube, the screen grid is placed between the control grid and the plate. What is the main function of the screen grid?

(A) To reduce the electron flow from the cathode.

(B) To control the capacitance between the control grid and the plate.

(C) To increase the overall amplification of the tube.

(D) To regulate the heater voltage.

(14) Which statement about a forward-biased LED is true?

(A) It will emit light.

(B) It will block the current flow in a circuit.

(C) It will amplify the input signal.

(D) It will dissipate a significant amount of heat.

(15) Which of these is not a technique for impedance matching in radio frequency circuits?

(A) Coaxial cable length.

(B) Bleeder resistors.

(C) LC Pi networks.

(D) Impedance-matching transformers.

(16) An amateur radio operator transmits with a peak voltage of 300 volts into a 50-ohm antenna. What is the peak envelope power (PEP) of the transmission?

(A) 450 watts.

(B) 900 watts.

(C) 1800 watts.

(D) 3600 watts.

(17) An audio amplifier has an output impedance of 4 ohms. What turns ratio should a transformer have to match this amplifier to an 8-ohm speaker?

(A) 1:$\sqrt{2}$.

(B) $\sqrt{2}$:1.

(C) 1:2.

(D) 2:1.

(18) Sarah recently acquired a vacuum tube amplifier and is tuning for maximum power to transmit on the 20-meter band. Which control adjustment should she be most cautious of to ensure she does not exceed the tube's plate current limitations?

(A) LOAD control.

(B) BAND switch.

(C) RECOUPLED control.

(D) MODE switch.

(19) An amateur radio operator measures the resistance of a component and requires a highly precise reading. Which type of multimeter will be the best choice in this scenario?

(A) Analog multimeter.

(B) Digital multimeter.

(C) Clamp meter.

(D) LCR meter.

(20) Which component is mainly used to shunt RF energy away from sensitive circuits to reduce interference?

(A) Resistor.

(B) Diode.

(C) Inductor.

(D) Capacitor.

(21) What type of modulation causes transmitted frequencies to extend above the carrier frequency?

(A) Single Sideband Upper Sideband.

(B) Frequency Modulation.

(C) Continuous Wave.

(D) Single Sideband Lower Sideband.

(22) Which statement about mobile HF stations is untrue?

(A) Mobile antennas usually have a narrower bandwidth than full-size antennas.

(B) A high-power HF transmitter should be connected directly to the car's battery with thick wires for optimal performance.

(C) A vehicle's fuel system can interfere with received HF signals.

(D) Power can be supplied to an HF transceiver from the cigarette lighter socket to save time.

(23) The K-index and A-index are used to quantify geomagnetic activity. Which of these statements accurately describes what they measure?

(A) The K-index measures the long-term average of geomagnetic activity, while the A-index measures short-term fluctuations.

(B) The K-index measures solar wind strength, while the A-index measures the density of charged particles in the ionosphere.

(C) The K-index measures short-term fluctuations in the Earth's magnetic field, while the A-index represents a daily average.

(D) The K-index measures the number of sunspots on the solar surface, while the A-index measures the intensity of solar flares.

(24) Which of these correctly lists the maximum single-hop distances for the layers of the ionosphere?

(A) D layer: 2500 miles, E layer: 1200 miles, and F2 layer: 180 miles.

(B) D layer: 180 miles, E layer: 1200 miles, and F2 layer: 2500 miles.

(C) D layer: 1200 miles, E layer: 2500 miles, and F2 layer: 180 miles.

(D) D layer: 2500 miles, E layer: 180 miles, and F2 layer: 1200 miles.

(25) During daylight hours, what frequency band will most likely experience significant signal attenuation due to D region absorption?

(A) 2 meters.

(B) 10 meters.

(C) 20 meters.

(D) 80 meters.

(26) An operator receives a Radioteletype (RTTY) signal, but the characters on their decoder screen are garbled and unreadable. Which action could fix the issue?

(A) Increase the SWR of the antenna.

(B) Reverse the polarity of the power supply.

(C) Check the decoder software for mark and space-frequency reversal.

(D) Adjust the squelch control on the receiver.

(27) Which statement about amateur radio contests is not true?

(A) Power limits and frequency restrictions must adhere to FCC regulations.

(B) Contest logs are usually submitted to sponsors for scoring and potential awards.

(C) New licensees are discouraged from participating.

(D) Contests can be a fun way to improve operating skills.

(28) In a CW signal report of 599C, what does the C indicate about the received signal?

(A) The signal is perfectly copied.

(B) The report is based on the S-meter.

(C) The signal has a chirpy quality.

(D) The signal has noticeable key clicks.

(29) An operator prepares to send a CQ call on a CW frequency that appears to be clear. Before transmitting, they want to ensure they are not interrupting another QSO. How do they check if the frequency is in use?

(A) Wait a few minutes, then transmit the CQ call if no activity is heard.

(B) Send QTH? followed by their call sign.

(C) Send QRL? followed by their call sign.

(D) Send a series of dots (dits) followed by their call sign.

(30) In an emergency situation on a ham radio frequency, what does it mean if a station repeatedly says "break"?

(A) They are testing their equipment.

(B) They are indicating a critical emergency.

(C) They are requesting a signal report.

(D) They are ending their transmission.

(31) To what type of transmission does the 10-watt PEP limit apply?

(A) Single sideband (SSB).

(B) Frequency modulation (FM).

(C) Spread spectrum.

(D) Amplitude modulation (AM).

(32) What is not a requirement for accreditation as a volunteer examiner (VE)?

(A) Hold a General Class or higher FCC-issued amateur radio license.

(B) Be at least eighteen years old.

(C) Be a US citizen.

(D) Obtain accreditation through a volunteer examiner coordinator (VEC).

(33) Which action is not mandatory before deploying a novel digital mode on amateur radio frequencies?

(A) Share technical details of the protocol publicly.

(B) Ensure the mode complies with existing band plans and regulations.

(C) Register the new mode with the International Amateur Radio Union (IARU).

(D) Test the mode extensively for compatibility and interference issues.

(34) Which statement accurately describes the power limitations for amateur radio stations?

(A) Beacon stations are limited to 100 watts PEP output.

(B) All amateur stations are limited to 100 watts PEP output.

(C) Technician Class licensees are limited to 200 watts PEP output on all HF bands.

(D) Novice Class licensees are limited to 5 watts PEP output on all bands.

(35) Which requirement is not necessary for amateur stations operating as secondary users?

(A) Avoid harmful interference to primary users.

(B) Accept interference from primary users.

(C) Cease operation when requested by the primary user.

(D) Obtain prior approval from the primary user before transmitting.

Test 5: Answers and Explanations

(1) (B) It must identify itself with a call sign.

For stations transmitting with a time-averaged power above 0.5 milliwatts, identification is required to comply with FCC rules, which mandate call sign identification for tracking and accountability purposes in regulated bands.

(2) (A) To create a low-resistance path for lightning current to dissipate into the ground.

Bonding all ground rods together in a lightning protection system is essential to create a low-resistance path for lightning current to dissipate into the ground. This helps to minimize the potential for electrical damage and fire caused by lightning strikes. Lightning protection systems are designed to safely intercept and redirect lightning strikes away from structures and equipment. Ground rods are conductive rods driven into the ground to provide a low-resistance path for electrical current to flow into the Earth.

(3) (B) Reduced power reaching the antenna for radiation.

Standing Wave Ratio (SWR) measures how efficiently power is transferred from a transmitter to an antenna through a transmission line, such as a coaxial cable. An ideal SWR of 1:1 signifies a perfect match, where all power is radiated by the antenna. However, a high SWR indicates a mismatch between the transmission line and the antenna. Depending on the degree of mismatch, a portion of the transmitted power is reflected back toward the transmitter. This reflection has several negative consequences:

- It reduces the effective radiated power (ERP), as the reflected power does not contribute to the intended signal.
- It increases losses (often as heat) in the transmission line due to the additional power traveling back and forth.
- The high voltage and current caused by the reflected power can damage the transmitter's final amplifier stage in extreme cases.

(4) (A) It requires a ground plane or radial system for efficient operation.

A quarter-wave vertical antenna is a monopole antenna, meaning it is half of a dipole and relies on a ground plane or radial system to reflect the signal and improve efficiency, particularly on lower HF bands.

(5) (D) The main lobe.

The main lobe of a directional antenna is the region of its radiation pattern where the radiated field strength is highest. The radiation pattern is a graphical representation of how an antenna radiates energy in different directions. The main lobe is the central area of focus for the antenna's transmitted or received signal, and it is usually the direction the antenna is pointed. The tip of the antenna might be physically the furthest point, but it is not necessarily where the maximum radiation occurs. Although areas of high current density contribute to radiation, the main lobe represents the combined effect of all radiating elements and is where the field strength is maximized. In Yagi antennas, the directors are passive elements that help shape the radiation pattern and increase directivity. They do not individually produce the maximum field strength.

(6) (C) It conserves horizontal space.

An inverted-V dipole is a type of dipole antenna where the two arms are angled upward instead of horizontally. This configuration conserves horizontal space compared to a traditional horizontal dipole. This makes it a popular choice for installations where space is limited, such as backyards or rooftops. The gain of an inverted-V dipole is generally comparable to a horizontal dipole at the same height. However, the front-to-back ratio of an inverted-V dipole is usually not as good as a horizontal dipole. A balun (balanced to unbalanced transformer) is often used with dipole antennas to ensure a balanced feedline and prevent unwanted radiation from the outer shield of coaxial cables. Although an inverted-V dipole can sometimes be fed with a coaxial cable without a balun, a balun is still recommended for optimal performance.

(7) (A) Flat-topping.

Flat-topping is the most likely consequence of excessive audio input levels in an AM transmission. As mentioned earlier, when the audio input levels are too high, the AM signal can exceed the carrier wave's amplitude limits. This causes the waveform's peaks to be clipped or flattened, a phenomenon known as flat-topping. Carrier shift refers to a change in the average power of the carrier wave, which is not directly caused by excessive audio input levels. Frequency drift is a gradual change in the carrier wave frequency. It is usually caused by temperature changes or component instability, not excessive audio input. Spurious emissions are unintended signals generated by a transmitter. They are not directly related to excessive audio input levels.

(8) (A) 184.5 Hz.

The frequency deviation (Δf) of a reactance-modulated oscillator in an FM transmitter can be calculated using the following formula $\Delta f = (\Delta F\_max) \div (f\_out \div f\_in)$, where:

- Δf = Frequency deviation (Hz).
- ΔF_max = Maximum deviation frequency (Hz).
- f_out = Output frequency of the FM transmitter (Hz).
- f_in = Input frequency of the reactance-modulated oscillator (Hz).

In the given problem:

- ΔF_max = 3 kHz = 3000 Hz.
- f_out = 145.17 MHz = 145.17×10^6 Hz.
- f_in = 8.93 MHz = 8.93×10^6 Hz.

Substitute the values into the formula.

- $\Delta f = 3000 \text{ Hz} \div (145.17 \times 10^6 \text{ Hz} \div 8.93 \times 10^6 \text{ Hz})$.
- $\Delta f = 184.5423 \text{ Hz} \approx 184.5 \text{ Hz}$.

Therefore, the frequency deviation of the 8.93 MHz oscillator is approximately 184.5 Hz.

(9) (D) By one or more vertical lines on either side of the signal.

On a waterfall display, overmodulation in an AM signal is usually indicated by one or more vertical lines on either side of the signal. A waterfall display is a spectrum analyzer display that shows the frequency spectrum over time, with the most recent data at the top and older data scrolling down like a waterfall. Overmodulation occurs when the amplitude of the modulating signal (like audio) exceeds the limits of the carrier wave and causes distortion in the transmitted signal. This distortion manifests as splatter, which spreads the signal's energy into adjacent frequencies. The vertical lines on the waterfall display represent this splatter and indicate that overmodulation has occurred.

(10) (B) Tapped inductor.

The symbol shown represents a tapped inductor with one or more additional connection points along its winding. The connection points allow the creation of multiple inductances within a single component. Inductors are passive electronic components that store energy in a magnetic field when current passes through them. They are often used in filters, tuned circuits, and power supplies to control the flow of alternating current (AC). A tapped inductor provides flexibility in circuit design by allowing the user to select different inductance values depending on where they connect the circuit. The dots or circles on the symbol usually indicate the tap points. A Zener diode operates in the reverse breakdown region to provide a stable reference voltage. Its symbol includes a line resembling a Z at the cathode end. An NPN transistor is a three-terminal device (base, emitter, and collector) that amplifies current. Its symbol is similar to a FET but has a straight line for the base terminal instead of an arrow and lacks the tap points seen on an inductor symbol. A transformer is a passive electrical device that transfers electrical energy between two or more circuits through mutual inductance. Its symbol usually consists of two coils (that represent the primary and secondary windings) with straight lines between them.

(11) (C) Shift register.

A shift register is a digital circuit that systematically moves data bits through a series of storage elements under the control of a clock signal. Each storage element in the shift register usually stores one bit of data. When the clock signal transitions, the data bits shift from one storage element to the next in a predetermined direction (left or right). A multiplexer (MUX) is a digital circuit that selects one of several input lines and forwards the selected input to a single output line. A demultiplexer (DEMUX) is a digital circuit that takes a single input line and distributes it to one of several output lines. A counter is a digital circuit that counts the number of clock pulses that have occurred.

(12) (A) Digital signal processors can operate on only one type of bandwidth.

Digital signal processors (DSPs) are versatile devices that can be programmed to operate on a wide range of bandwidths, from audio frequencies to RF and beyond. They can perform various signal-processing tasks, such as filtering, modulation, demodulation, and encoding/decoding. In I/Q modulation, the phase relationship between the in-phase (I) and quadrature (Q) signals is ninety degrees. This is a fundamental property of I/Q modulation, a technique used to represent signals in the complex plane for efficient transmission and reception. The bandwidth of the demodulator affects its ability to capture and process the received signal, which influences the receiver's sensitivity. SDRs are radios where much signal processing is done with software rather than hardware. They are flexible and can be programmed to perform various functions, such as filtering, modulation, demodulation, and detection.

(13) (B) To control the capacitance between the control grid and the plate.

In a tetrode vacuum tube, the screen grid is a mesh-like structure between the control grid and the plate (anode). Its main function is to reduce the capacitance between these two electrodes. Capacitance is the ability of a system to store an electric charge. The capacitance between the control grid and plate in a vacuum tube can cause instability and oscillation due to feedback effects. By reducing this inter-electrode capacitance, the screen grid helps to stabilize the tube's operation, especially at higher frequencies. This ensures that the tube functions as a linear amplifier and avoids unwanted oscillations. Although the screen grid also has a secondary effect of accelerating electrons toward the plate, its main purpose is to minimize the capacitance between the control grid and the plate.

(14) (A) It will emit light.

A forward-biased LED (Light Emitting Diode) will emit light. A diode is a two-terminal electronic component that allows current to flow in one direction. An LED is a specialized diode that emits light when an electrical current passes through it in the forward direction. Forward biasing refers to applying a voltage across the diode such that the positive terminal is connected to the anode and the negative terminal is connected to the cathode. Electrons and electron holes recombine within the semiconductor material when an LED is forward-biased. This process releases energy through photons perceived as light. The color of the emitted light depends on the specific semiconductor material used in the LED.

(15) (B) Bleeder resistors.

Bleeder resistors are used in power supplies to safely discharge capacitors when the power is turned off and protect users from electric shock. They are not used for impedance matching. Impedance matching ensures that the impedance of a source (like a transmitter) is equal to that of a load (like an antenna) for maximum power transfer and minimum signal reflection. Impedance mismatch can cause signal loss, reduced efficiency, and potential damage to equipment. The length of a coaxial cable can be used as a transmission line to match impedances. LC Pi networks are filter networks consisting of two capacitors and an inductor arranged in a pi (π) configuration. They are commonly used to match impedance in RF circuits by adjusting the values of the components to create a resonant circuit that matches the desired impedance. Impedance-matching transformers change the ratio of voltage to current in the circuit to transform the impedance seen by the source or load. They are often used to match the impedance of an antenna to the impedance of a transmitter or receiver.

(16) (B) 900 watts.

PEP is the maximum power the transmitter can deliver during a transmission cycle. The formula to calculate PEP is PEP = $Vrms^2 \div R$, where PEP is the peak envelope power in watts (W), Vrms is the root-mean-square voltage in volts (V), and R is the resistance (antenna impedance) in ohms (Ω). From the information in the question, Vpeak (peak voltage) = 300 V and R (antenna impedance) = 50 Ω. Use the following formula to convert the peak voltage to RMS voltage: Vrms = Vpeak × 0.707.

Substitute the values.
Vrms = 300 × 0.707 = 212.1 V

Use these values in the PEP formula.
$(212.1\ V)^2 \div 50\ \Omega = 899.7282\ W \approx 900\ W$.

Therefore, the peak envelope power (PEP) is approximately 900 watts.

(17) (B) $\sqrt{2}$:1.

The formula to calculate a transformer's turns ratio (TR) is TR = √(Impedance Ratio), where Impedance Ratio = Feed Line Impedance ÷ Transmitter Impedance. In the question, feed line impedance (speaker) = 8 ohms and transmitter impedance (amplifier) = 4 ohms.

> Substitute the values into the Impedance Ratio formula.
> Impedance Ratio = 8 ohms ÷ 4 ohms = 2:1
>
> Use this value in the TR formula.
> *TR* = √2:1.

Therefore, the transformer needs a turn ratio of √2:1 to match the 4-ohm amplifier to the 8-ohm speaker. This means the primary winding (connected to the amplifier) should have √2:1 times more turns than the secondary winding (connected to the speaker).

(18) (A) LOAD control.

The LOAD or COUPLING control on a vacuum tube amplifier is used to match the impedance of the amplifier's output to the antenna. When the load is matched correctly, the maximum amount of power is transferred from the amplifier to the antenna, and the plate current in the output tube is within safe limits. However, if the load control is misadjusted, it can cause a high Standing Wave Ratio (SWR), which means that some of the power is reflected back into the amplifier instead of being radiated. This reflected power can cause excessive plate current and potentially damage the tube. The band switch selects the operating frequency band, while the mode switch selects the operating mode (such as CW, SSB, and AM). There is no RECOUPLED control on a vacuum tube amplifier.

(19) (B) Digital multimeter.

A digital multimeter (DMM) is the most suitable tool for measuring resistance with high precision. DMMs provide higher resolution and accuracy than analog multimeters due to their digital display and internal circuitry. Depending on the model and range setting, they can measure resistance to several decimal places. Analog multimeters use a needle and scale to display readings, which are less precise than digital displays. Clamp meters are designed to measure high AC or DC currents without breaking the circuit. They are not ideal for measuring resistance, especially with high precision. An LCR meter is a specialized instrument used to measure inductance (L), capacitance (C), and resistance (R). Although it can measure resistance, it may not be as versatile or user-friendly as a DMM for general-purpose measurements.

(20) (D) Capacitor.

A bypass capacitor is used to shunt RF energy away from sensitive circuits to reduce interference. They are connected in parallel with a circuit's power supply or signal lines and act as a low-impedance path for high-frequency signals like RF energy. Capacitors provide very low impedance to high-frequency signals due to their ability to store and discharge electrical charge rapidly. When RF energy encounters a bypass capacitor, it sees a much easier path to ground than through the circuit's components. The RF energy is shunted or diverted through the capacitor and into the ground. Thus, the capacitor prevents RF energy from entering the sensitive circuit and causing interference. This maintains the integrity of the signal within the circuit and protects it from external noise.

(21) (A) Single Sideband Upper Sideband.

In Upper Sideband modulation, the transmitted frequencies are located above the carrier frequency. USB is commonly used on higher amateur radio bands, usually above 10 MHz, whereas Lower Sideband is used on lower bands.

(22) (D) Power can be supplied to an HF transceiver from the cigarette lighter socket to save time.

Although it might seem convenient to power an HF transceiver from the cigarette lighter socket (12 V accessory socket), this is not recommended, especially for high-power transmitters. HF transceivers draw significant current during transmission, especially those operating at higher power levels. The cigarette lighter socket is not designed to handle such high currents and may overheat or even melt. This can even damage the radio or the vehicle's electrical system. The wiring used for the cigarette lighter socket is often thin and not intended for high-current applications. This can cause a significant voltage drop, especially during transmission, which can degrade the transceiver's performance. High-power HF transceivers in mobile installations should be connected directly to the vehicle's battery using thick, low-resistance cables. This ensures adequate current delivery and minimizes voltage drop for optimal performance and reliability of the radio.

(23) (C) The K-index measures short-term fluctuations in the Earth's magnetic field, while the A-index represents a daily average.

The K-index and A-index are geomagnetic indices that characterize the magnitude of geomagnetic storms and their potential impact on radio communication. K-index is a three-hour quasi-logarithmic local index that measures the disturbance level in the horizontal component of Earth's magnetic field at a particular location. It ranges from 0 (quiet) to 9 (severe storm). The K-index is a good indicator of short-term fluctuations in geomagnetic activity. A-index is a daily average of the K-index. It provides a more general picture of geomagnetic activity over a twenty-four-hour period. It ranges from 0 to 400, with higher values indicating more intense geomagnetic storms.

(24) (B) D layer: 180 miles, E layer: 1200 miles, F2 layer: 2500 miles.

The ionosphere is divided into several layers based on altitude and ionization characteristics. Each layer has a different maximum single-hop distance, the farthest a signal can travel by being reflected off that layer once. The D layer (50-90 km) is the lowest and least ionized. It mainly absorbs low-frequency signals. Its maximum single-hop distance is approximately 180 miles (300 km). The E layer (90-150 km) is more ionized than the D layer and can support short-distance skywave propagation. Its maximum single-hop distance is around 1200 miles (2000 km). The F2 layer (200-400 km) is the highest and most ionized layer that facilitates long-distance HF communication. Its maximum single-hop distance is about 2500 miles (4000 km).

(25) (D) 80 meters.

The D region is the lowest layer of the ionosphere and is the most absorbent of HF radio waves, especially during daylight hours. This absorption is stronger for lower frequencies. Therefore, the 80-meter band, with its lower frequencies than 2, 10, and 20 meters, is most likely to experience significant signal attenuation due to D region absorption during the day.

(26) (C) Check the decoder software for mark and space-frequency reversal.

RTTY (Radioteletype) signals are transmitted using two audio tones, often called mark and space. These tones represent the binary ones and zeros that comprise the digital data. If the mark and space frequencies are reversed in the decoder software, the characters appear garbled and unreadable. This common issue can be easily fixed by adjusting the mark and space settings in the software to match the transmitted signal. SWR (Standing Wave Ratio) measures how well an antenna is matched to a transmitter. Although a high SWR can indicate a problem with the antenna system, it will not directly cause garbled characters in RTTY decoding. Reversing the polarity of the power supply will likely damage the radio equipment and is not a solution for RTTY decoding issues. Ssquelch is a feature that helps reduce background noise when no signal is present. Adjusting squelch might improve the audio quality but will not fix the reversed mark and space frequencies problem.

(27) (C) New licensees are discouraged from participating.

Amateur radio contests are competitive events where operators try to make as many contacts as possible within a specific period and under a set of rules. These contests are not just for experienced operators; they are open to all licensed radio amateurs, regardless of experience level. In fact, new licensees are encouraged to participate as contests provide a fun and engaging way to improve operating skills, learn about different modes and bands, and make contacts with operators from around the world. All amateur radio activities, such as contests, must comply with the FCC's rules and regulations (or the equivalent regulatory body in other countries). Contest logs are detailed records of the contacts made during the contest. They contain information such as call signs, signal reports, and times. These logs are submitted to the contest sponsors, who use them to calculate scores and determine winners. Many contests give awards or certificates to recognize the achievements of participants. Contests provide a unique environment for practicing and honing operating skills.

(28) (C) The signal has a chirpy quality.

In a CW (Morse code) signal report, the optional C following the RST (Readability, Strength, Tone) code indicates that the received signal has a chirpy quality. A chirpy signal is one where the transmitted signal's frequency drifts slightly during the transmission of each dot or dash and causes a noticeable change in pitch. Various factors, such as power supply instability or poor keying technique, can cause this. A perfectly copied signal will usually be indicated by a 599 signal report alone, where 5 refers to perfect readability, the first 9 means signal strength, and the second 9 represents perfect tone. Although the signal might be perfectly copied, the C indicates that the signal is nevertheless chirpy or unstable. The receiver's S-meter (signal strength meter) can give a general idea of signal strength. However, the RST (Readability, Strength, Tone) report is a more subjective assessment based on the operator's perception. The C is not related to the S-meter reading. Key clicks are sharp transients caused by the opening and closing of the telegraph key. Clicks are indicated by a different code (K) in the signal report.

(29) (C) Send QRL? followed by their call sign.

In amateur radio, Q signals are three-letter codes used for concise communication. QRL? is a specific Q signal that translates to "Is this frequency in use?" Before transmitting on a seemingly clear frequency, send QRL? followed by a call sign. This is a courtesy to other operators who may be listening but not transmitting. If the frequency is in use, the other station will respond with QRL, which means, "Yes, this frequency is in use." The operator may proceed with their CQ call or other transmission if the frequency is clear. QTH? is a Q signal that means "What is your location?" Sending this will not determine if the frequency is in use.

(30) (B) They are indicating a critical emergency.

When "break" is repeated on a ham radio frequency, it signals an urgent or emergency situation and alerts other stations to stop transmissions to allow the emergency station priority.

(31) (C) Spread spectrum.

The 10-watt PEP (Peak Envelope Power) limit applies specifically to spread spectrum transmissions on certain amateur radio bands. This restriction limits potential interference with other services due to the wider bandwidth used by spread spectrum signals. Other modes like SSB, FM, and AM might have different power limits depending on the specific frequency and regulations. Spread spectrum is a modulation technique that spreads the signal over a wider bandwidth, which makes it more resistant to interference and jamming. However, this wider bandwidth may increase its chance of interference with other services, hence the power restriction.

(32) (C) Be a US citizen.

There is no requirement for volunteer examiners (VEs) to be US citizens. VEs must hold a General Class or higher license, be at least eighteen years old, and obtain accreditation through a volunteer examiner coordinator (VEC). This allows a diverse pool of qualified individuals to contribute to the amateur radio community by administering exams and helping others obtain their licenses.

(33) (C) Register the new mode with the International Amateur Radio Union (IARU).

To deploy a new digital mode, operators must share technical details, test the mode, and ensure compliance. However, it is not mandatory to register a new mode with the IARU. The IARU is a non-governmental organization that represents the interests of amateur radio operators worldwide, but it does not have regulatory authority over new digital modes.

(34) (A) Beacon stations are limited to 100 watts PEP output.

Beacon stations are automated transmitters used to assess propagation conditions. They are limited to a maximum power output of 100 watts PEP (Peak Envelope Power). Not all amateur stations are limited to 100 watts PEP output. The power limit varies depending on the license class and the frequency band. Technician Class licensees are limited to 200 watts PEP output on HF bands, not all bands.

(35) (D) Obtain prior approval from the primary user before transmitting.

Amateur radio operators often act as secondary users on some frequencies. That is, they share those frequencies with primary users who have priority. Avoiding harmful interference to the primary user is paramount as a secondary user. Secondary users must also accommodate interference and cease operation upon request. However, prior approval from the primary user before transmitting is not a requirement.

Test 6: Questions

(1) Which statement regarding repeater operation on the 10-meter band is true?

(A) Repeater operation is completely prohibited on the 10-meter band.

(B) Repeater operation is permissible anywhere within the 10-meter band.

(C) Repeater operation is restricted to frequencies above 29.5 MHz.

(D) Repeater operation is only allowed during daylight hours.

(2) Amateur radio communication with foreign stations is permissible as long as:

(A) The foreign station's administration does not object.

(B) The communication is brief and non-commercial.

(C) The foreign station is located in North America.

(D) The communication occurs on designated frequencies.

(3) Which of these statements about PEP is not true?

(A) It is measured in watts.

(B) It is higher than the average power output.

(C) It is the same as the RMS.

(D) It is regulated by the FCC.

(4) How long can an operator hold a Certificate of Successful Completion of Examination (CSCE) for an exam element before having to retake that exam element?

(A) One month.

(B) Three months.

(C) 180 days.

(D) One year.

(5) In amateur radio, what constitutes a third-party message?

(A) Any message that does not originate from the control operator.

(B) A message intended for someone other than the control operator or station licensee.

(C) A message that contains commercial or political content.

(D) A message relayed through a repeater or digital network.

(6) In what scenario will an amateur radio operator most likely prefer VOX (voice-operated transmit) over PTT (push-to-talk) for SSB communication?

(A) When they seek the highest audio fidelity.

(B) When they must operate the radio without using their hands.

(C) When they attempt to minimize power consumption.

(D) When the environment is noisy.

(7) Which of the statements below is not correct?

(A) A minimum of 150-500 Hz separation is needed between CW signals to avoid disruption.

(B) Only FCC-licensed radio amateurs can control stations in RACES operations.

(C) The voluntary band plan is the main resource for choosing frequencies.

(D) Nets must adhere to only one central frequency during interferences.

(8) What is the correct way for an operator to tell a transmitting station that they are prepared to receive a transmission in CW?

(A) Send a QRL? signal.

(B) Send a QSL signal.

(C) Send a QRV signal.

(D) Send a QSY signal.

(9) A ham radio operator attempts to contact a station on the other side of the world. She initially points her directional antenna directly toward the station's location but cannot establish a connection. Out of frustration, she randomly turns her antenna in the opposite direction and unexpectedly receives a strong signal from the target station. What phenomenon most likely explains this successful reception?

(A) Atmospheric skip.

(B) Sporadic E propagation.

(C) Long-path propagation.

(D) Tropospheric ducting.

(10) Which organization uses amateur radio to transmit data at high speeds during emergencies?

(A) RACES.

(B) Winlink.

(C) AREDN.

(D) SKYWARN.

(11) What phenomena is correctly matched with its usual duration or time of effect?

(A) Solar flare takes fifty to sixty days to affect communications.

(B) Coronal mass ejection affects radio within fifteen hours to several days.

(C) Sunspot cycle occurs every eleven months.

(D) Solar wind is produced every five hours.

(12) Which statement about HF radio wave propagation is false?

(A) Signal attenuation occurs when the frequency is lower than the LUF.

(B) Increased solar activity results in a higher MUF for a given path.

(C) Lower HF frequencies are more susceptible to static noise during periods of high solar radiation.

(D) Communication over HF is improved when the LUF surpasses the MUF.

(13) Near Vertical Incidence Skywave (NVIS) propagation is most suitable for which type of communication?

(A) Communication over a few hundred kilometers.

(B) Long-distance international communication.

(C) Point-to-point communication between fixed stations.

(D) Communication with satellites.

(14) In which scenario will the dual VFO feature on a transceiver be most beneficial?

(A) To monitor two different bands simultaneously.

(B) To operate in split frequency mode during a contest.

(C) To improve receiver selectivity.

(D) To reduce transmitter power output.

(15) What is an antenna system's ideal Standing Wave Ratio (SWR)?

(A) 3:1.

(B) 1:1.

(C) 2:1.

(D) 10:1.

(16) Ferrite chokes are commonly used to mitigate which type of electrical noise in cables?

(A) Differential-mode noise.

(B) White noise.

(C) Thermal noise.

(D) Common-mode noise.

(17) If a radio receiver's S-meter reading changes from S5 to S6, which of the statements regarding the change in signal strength and received power output is correct?

(A) The signal strength increases by 6 dB, and the received power output quadruples.

(B) The signal strength increases by 1 dB, and the received power output doubles.

(C) The signal strength cannot be determined from the S-meter change alone, but the received power output quadruples.

(D) The signal strength increases by 6 dB, and the received power output doubles.

(18) Which of these statements about the functions of solar charging system components is not correct?

(A) The solar panel converts sunlight into direct current (DC) electricity.

(B) The charge controller regulates the voltage and current to prevent battery damage.

(C) The inverter converts DC electricity from the battery into alternating current (AC) for household use.

(D) The blocking diode blocks excess current from reaching the batteries.

(19) Which of these statements about reactance is true?

(A) It is represented by the symbol Z and is measured in Ohms (Ω).

(B) It is represented by the symbol R and is measured in Ohms (Ω).

(C) It is represented by the symbol X and is measured in Ohms (Ω).

(D) It is represented by the symbol X and is measured in Siemens (S).

(20) What AC voltage value will produce the same amount of heat in a resistor as a 10 V DC source?

(A) 10 V peak.

(B) 10 V peak-to-peak.

(C) 10 V RMS.

(D) 14.14 V RMS.

(21) What is the main function of a capacitor in an AC circuit?

(A) To store energy in a magnetic field.

(B) To oppose changes in voltage.

(C) To convert AC to DC.

(D) To increase current flow.

(22) What is the most likely outcome as the operational frequency of a circuit containing an inductor increases and surpasses its self-resonant frequency?

(A) The inductor's reactance will decrease.

(B) The inductor will begin to act more like a capacitor.

(C) The inductor's current will increase linearly.

(D) The inductor will short circuit.

(23) Which of these best describes how a ferrite choke works?

(A) It acts as a low-pass filter and enhances high-frequency RF noise.

(B) It generates a magnetic field that cancels out the RF noise.

(C) It increases the cable's resistance to RF noise and forces it to flow through the core.

(D) It converts the common-mode RF noise into differential-mode noise, which is easier to filter.

(24) What is the main control element in a Field Effect Transistor?

(A) Base.

(B) Emitter.

(C) Gate.

(D) Collector.

(25) Which of these statements is not true?

(A) The resonant frequency of an LC oscillator is directly proportional to the square root of the product of inductance (L) and capacitance (C).

(B) Linear amplifiers preserve the input waveform shape.

(C) Frequency Modulation (FM) is well-suited for Class C amplifiers.

(D) The distortion introduced by Class C amplifiers makes them unsuitable for AM schemes.

(26) Which of these is true about low-pass filters?

(A) They completely block all frequencies above the cutoff frequency.

(B) They amplify signals below the cutoff frequency.

(C) The output signal power is less than half the input power at the cutoff frequency.

(D) They usually remove low-frequency noise from audio signals.

(27) What is Frequency Shift Keying?

(A) A method of varying the frequency of a carrier to transmit digital data.

(B) A technique to modulate amplitude for voice transmission.

(C) A process to control phase shift in a carrier signal.

(D) A form of digital modulation that changes the carrier's bandwidth.

(28) If an FM signal has a maximum frequency deviation of 10 kHz and is modulated by a signal with a frequency of 4 kHz, what is the resulting bandwidth of the FM signal?

(A) 4 kHz.

(B) 10 kHz.

(C) 14 kHz.

(D) 28 kHz.

(29) In Frequency Shift Keying (FSK), what do the terms *mark* and *space* represent?

(A) The amplitude levels of the carrier signal.

(B) The phase difference between two carrier signals.

(C) The two distinct frequencies used to represent binary 1 and 0.

(D) The time duration of each bit in the digital signal.

(30) In what situation does coaxial cable attenuation decrease?

(A) As the frequency of the signal increases.

(B) During decreases in signal frequency.

(C) When the Standing Wave Ratio increases.

(D) As the characteristic impedance stabilizes.

(31) What is the typical impedance of a half-wave dipole antenna in free space?

(A) 50 ohms.

(B) 75 ohms.

(C) 300 ohms.

(D) 450 ohms.

(32) A Yagi antenna is designed with a feed point impedance of 30 ohms. The radio amateur decides to use a beta match to match this impedance to a standard 50-ohm coaxial cable. Which statement accurately describes the beta match?

(A) It places a quarter-wavelength transformer at the feed point.

(B) It uses a variable capacitor for tuning.

(C) It consists of a shorted transmission line stub placed at the feed point.

(D) It is a type of balanced feedline used to connect the antenna to the radio.

(33) Which of these statements is incorrect?

(A) Harmonic rejection prevents interference with other radio users.

(B) Multiband antennas have great harmonic rejection.

(C) An inverted V dipole has a single central support.

(D) Log-periodic antennas have narrow bandwidths.

(34) What is one method for minimizing exposure to direct RF energy?

(A) Decreasing the antenna height.

(B) Increasing transmission power.

(C) Increasing distance from the source.

(D) Reducing the frequency of transmission.

(35) What is the main purpose of the National Electrical Code (NEC)?

(A) To standardize electrical equipment manufacturers' warranties.

(B) To establish guidelines for radio frequency emissions from electrical devices.

(C) To safeguard people and property from electrical hazards by providing wiring and installation standards.

(D) To regulate the maximum allowable voltage in residential electrical systems.

Test 6: Answers and Explanations

(1) (C) Repeater operation is restricted to frequencies above 29.5 MHz.

Repeater operation on the 10-meter band is not entirely prohibited, but it is subject to restrictions. Repeaters can only operate on frequencies above 29.5 MHz within the 10-meter band. This limitation is in place to ensure efficient use of the band and minimize potential interference with other users, especially those using other modes or services in the lower portion of the band.

(2) (A) The foreign station's administration does not object.

Amateur radio communication with foreign stations is generally permissible as long as the administration of the foreign country does not object. There may be specific restrictions or agreements in place depending on the countries involved, but the primary condition is that the foreign administration allows such communication. Although most amateur radio communication is non-commercial, it is not required to be brief. Even though some bands may restrict international communication, there are no designated frequencies specifically for this purpose.

(3) (C) It is the same as the RMS.

PEP stands for Peak Envelope Power and is the maximum power of a radio signal during a single RF cycle. It is measured in watts and is regulated by the FCC. PEP is higher than the average power output, especially in modes like SSB, where the power varies. RMS (Root Mean Square) measures the effective voltage of an AC waveform, which is not the same as PEP.

(4) (D) One year.

A Certificate of Successful Completion of Examination (CSCE) is valid for one year. If an operator does not upgrade their license or retake the exam within that year, they will need to retake the exam element to get credit for it again. This ensures that operators remain current with their knowledge and skills and encourages them to progress to higher license classes.

(5) (B) A message intended for someone other than the control operator or station licensee.

In amateur radio, a third-party message is any message not intended for the person operating the radio station (the control operator) or the person who holds the license for the station. This includes messages relayed to the licensee's family, friends, or acquaintances. The content of the message (such as commercial, political, or personal) or the method of transmission (such as a repeater or digital network) does not define it as third-party. The fundamental factor is the intended recipient.

(6) (B) When they must operate the radio without using their hands.

VOX (Voice Operated Transmit) is a feature in many amateur radios that facilitates hands-free operation. When VOX is enabled, the radio automatically switches to transmit mode when it detects audio input from the microphone. This can be useful when the operator's hands are occupied with other tasks, such as adjusting equipment, writing down information, or driving a vehicle. PTT (Push-to-Talk), on the other hand, requires the operator to press and hold a button while speaking. Although PTT is still widely used, VOX provides a convenient alternative in certain situations.

(7) (D) Nets must adhere to only one central frequency during interferences.

Nets are organized on-air meetings of amateur radio operators, usually with a specific purpose or topic. Although nets often operate on a designated frequency, they are not restricted to using only one frequency, especially in the face of interference. If interference is encountered, the net control station (NCS) may move the net to an alternate frequency to ensure clear communication and maintain the net flow. This flexibility helps maintain communication during emergencies or other situations where interference may occur. CW (Morse code) signals have a bandwidth, and if stations transmit too close to each other, their signals will overlap and become difficult to distinguish. The recommended 150-500 Hz separation ensures clear reception of both signals. RACES (Radio Amateur Civil Emergency Service) is a program that allows amateur radio operators to assist in emergencies. However, only licensed amateur radio operators can control RACES stations to ensure proper operation and regulation compliance. The voluntary band plan is a gentleman's agreement among amateur radio operators that outlines recommended frequency ranges for different modes of communication and activities. It helps operators choose appropriate frequencies to minimize interference and promote efficient spectrum use.

(8) (C) Send a QRV signal.

QRV means "I am ready." Sending this signal indicates to the transmitting station that the operator is prepared to receive a transmission. QRL? means "Is this frequency in use?" and is used to inquire about the availability of a frequency before initiating a transmission. QSL means "I acknowledge receipt" and is used to confirm that a message or transmission has been received and understood. QSY means "Shall I change to another frequency?" It is used to request a change in operating frequency, usually due to interference or other issues.

(9) (C) Long-path propagation.

In amateur radio, communication often relies on signals bouncing off the ionosphere to reach distant locations. The shortest path between two points on Earth is a great circle route. However, sometimes signals can take a longer path, known as long-path propagation. This occurs when signals travel more than halfway around the globe and go "the long way around" to reach the receiving station. Several factors can contribute to long-path propagation, such as the time of day, season, solar activity, and the specific frequencies being used. Atmospheric skip refers to the general phenomenon of HF signals bouncing off the ionosphere to enable long-distance communication. However, it does not explain the signal being received from the opposite direction. Sporadic E propagation is a short-lived phenomenon. Clouds of dense ionization in the E layer of the ionosphere can reflect signals back to Earth and enable unusual long-distance contacts, often on VHF frequencies. Tropospheric ducting occurs when temperature and humidity variations create a duct in the troposphere (the lowest layer of the atmosphere). Ducting traps and guides radio waves over extended distances.

(10) (C) AREDN.

AREDN (Amateur Radio Emergency Data Network) uses amateur radio to transmit data at high speeds during emergencies and community events. AREDN is a mesh network project where multiple nodes work together to create a robust and resilient network for data communication. This can be extremely valuable when traditional internet infrastructure is unavailable or compromised. RACES (Radio Amateur Civil Emergency Service) is a program that organizes amateur radio operators to provide communication support during emergencies. RACES may use digital modes for communication, but its central focus is on voice communication and message handling, not high-speed data services. Winlink is mainly a global radio email system designed to send and receive email messages over radio frequencies when internet access is limited. It can handle some data transfer but is not optimized for high-speed data like AREDN. SKYWARN is a program that trains amateur radio operators to observe and report severe weather conditions to the National Weather Service. It is used to monitor weather and public safety but is not focused on providing high-speed data services.

(11) (B) Coronal mass ejection affects radio within fifteen hours to several days.

Coronal mass ejections (CMEs) are large plasma and magnetic field expulsions from the Sun's corona. The time it takes for a CME to reach Earth and affect radio communication depends on its speed. Faster CMEs can arrive in as little as fifteen to eighteen hours, while slower ones may take several days. Solar flares affect radio communications almost immediately, usually within eight minutes, as they emit electromagnetic radiation that travels at the speed of light. The effects on the ionosphere can last for hours or even days, but the initial impact is nearly instantaneous. The sunspot cycle is a roughly eleven-year cycle during which the number of sunspots on the

Sun's surface increases and decreases. This cycle is linked to the Sun's magnetic activity and can affect radio propagation, but it does not occur every eleven months. The solar wind is a continuous stream of charged particles from the Sun's corona. It is not a periodic event that occurs every five hours. Although the solar wind's intensity can vary, it is always present and flowing outward from the Sun.

(12) (D) Communication over HF is improved when the LUF surpasses the MUF.

The Lowest Usable Frequency (LUF) is the minimum frequency required for skywave propagation between two points at a given time. It is determined by factors such as the distance between the stations, the time of day, and the absorption characteristics of the ionosphere. The Maximum Usable Frequency (MUF) is the highest frequency that can be refracted back to Earth. For successful skywave propagation, the operating frequency must be between the LUF and the MUF. If the LUF surpasses the MUF, there will be a radio blackout. In this case, no frequency range is suitable for skywave propagation, and communication will be impossible.

(13) (A) Communication over a few hundred kilometers.

Near Vertical Incidence Skywave (NVIS) propagation is a technique where HF radio waves are intentionally transmitted at high angles to be reflected almost vertically by the ionosphere. This results in short skip distances, which usually cover a few hundred kilometers. NVIS is useful for local and regional communication, such as broadcasting to specific areas, emergency communications, or providing coverage in mountainous terrain where line-of-sight communication is difficult.

(14) (B) To operate in split frequency mode during a contest.

Dual VFOs (Variable Frequency Oscillators) are a feature in many transceivers that allow the operator to have two independent receive and transmit frequencies. This is useful in contest scenarios, where operators often need to transmit on one frequency and listen for replies on another. Some transceivers allow the monitoring of two frequencies within the same band. However, simultaneous monitoring of two different bands often requires two separate receivers. Receiver selectivity is determined by the bandwidth and characteristics of the filters in the intermediate frequency (IF) stage, not VFOs. Transmitter power output is controlled by the power setting on the transceiver, not the VFOs.

(15) (B) 1:1.

SWR (Standing Wave Ratio) measures how well an antenna is matched to its transmission line. A 1:1 SWR indicates a perfect match. That is, the antenna radiates all transmitted power, and none is reflected back down the transmission line. This is the ideal scenario for maximum efficiency and optimal performance. 2:1 and 3:1 indicate a mismatch, where some of the transmitted power is reflected back toward the transmitter. Even though a small amount of reflected power is usually tolerable, higher SWR values can reduce efficiency, decrease signal strength, and damage the transmitter. 10:1 represents a severe mismatch, where a significant portion of the transmitted power is reflected back. This can cause serious damage to the transmitter if not corrected.

(16) (D) Common-mode noise.

Ferrite chokes are passive components that consist of ferrite material wrapped around a cable or formed into a bead-like structure. They effectively suppress common-mode noise, which is a type of electrical noise that appears on both conductors of a cable with the same polarity. This means the noise currents flow in the same direction on both wires. Ferrite chokes increase the impedance (resistance) of the cable to high-frequency signals, like common-mode noise. This is due to the ferrite material's properties, which cause it to dissipate the energy of the high-frequency noise as heat. The common-mode noise currents are blocked or attenuated. At the same time, the desired signal, which usually has a lower frequency and different impedance characteristics, is allowed to pass through with minimal attenuation.

(17) (A) The signal strength increases by 6 dB, and the received power output quadruples.

The S-meter on a radio receiver indicates the relative strength of the received signal, measured in S-units. Each S-unit represents a 6 dB (decibel) change in signal strength. The relationship between decibels (dB) and power is logarithmic, not linear. A 6 dB increase means the power has quadrupled, a 3 dB increase means the power has doubled, and a 10 dB increase means the power has increased tenfold. Conversely, a 3 dB decrease in signal strength means the power has been halved, and a 6 dB decrease means the power has been quartered. Therefore, a change from S5 to S6 on the S-meter represents a 6 dB increase in signal strength, which translates to a quadrupling of the received power output.

(18) (D) The blocking diode blocks excess current from reaching the batteries.

Solar panels convert sunlight into direct current (DC) electricity. The energy generated by the solar panels is stored in batteries for use when sunlight is unavailable, such as at night or during cloudy days. During periods of low or no sunlight, the voltage generated by the solar panel drops below the battery voltage. Without a blocking diode, the battery will try to discharge through the solar panel and act as a power source for the panel instead of being charged by it. This reverse current flow can damage the solar panel and reduce the battery's lifespan. The blocking diode acts as a one-way valve that allows current to flow from the solar panel to the battery when there is sufficient sunlight. It also prevents current from flowing in the opposite direction when sunlight is insufficient. A charge controller regulates voltage and current flow from the solar panel to the battery and prevents overcharging. The inverter converts the DC electricity stored in the battery into alternating current (AC) electricity.

(19) (C) It is represented by the symbol X and is measured in Ohms (Ω).

Reactance (X) is the opposition to the flow of alternating current (AC) caused by either inductance or capacitance in a circuit. It is a component of impedance (Z), which represents the total opposition to AC current flow. Reactance is measured in ohms (Ω) and is represented by the symbol X. Inductive reactance (X_L) is the opposition to AC caused by an inductor (a coil of wire), and it increases with increasing frequency. Capacitive reactance (X_C) is the opposition to AC caused by a capacitor (a device that stores electric charge), and it decreases with increasing frequency. The symbol Z represents impedance, while the symbol R represents resistance. Siemens is the unit of measurement for admittance (Y), which is the reciprocal of impedance ($1 \div Z$) and represents the ease with which AC current flows through a circuit.

(20) (C) 10 V RMS.

DC voltage and current are constant and do not change over time. Therefore, their average value equals their constant value, and there is no need for the RMS calculation. The average value of a DC signal accurately represents its power and energy delivery capability. On the other hand, RMS (Root Mean Square) is used in AC circuits because it provides a meaningful way to quantify the effective value of a constantly changing voltage or current. In AC circuits, the voltage and current levels oscillate between positive and negative values and pass through zero multiple times per second. As a result, the average value of an AC waveform over a complete cycle is zero, which does not provide a useful measure of its power or energy delivery capability. RMS accounts for the entire waveform (including positive and negative peaks). It is calculated by squaring the instantaneous values of the waveform, taking the average of those squares, and then calculating the square root of that average. This mathematical process weights the peaks of the waveform and gives it a value more representative of the AC signal's overall power and heating effect.

(21) (B) To oppose changes in voltage.

In an AC circuit, a capacitor stores energy in an electric field and opposes changes in voltage by charging and discharging in response to the voltage changes.

(22) (B) The inductor will begin to act more like a capacitor.

Due to its inherent capacitance, every inductor has a self-resonant frequency (SRF). This capacitance arises from the proximity of the wire turns in the coil, which creates small capacitive effects between them. At frequencies below the SRF, the inductor behaves primarily as an inductor, with its reactance (opposition to AC) increasing with frequency. However, as the operational frequency approaches and surpasses the SRF, capacitive reactance begins to dominate over inductive reactance. This is because capacitive reactance decreases with increasing frequency while inductive reactance increases. As a result, the inductor starts to behave more like a capacitor, with its overall impedance decreasing as the frequency increases beyond the SRF.

(23) (C) It increases the cable's resistance to RF noise and forces it to flow through the core.

A ferrite choke increases the impedance of a cable to high-frequency RF (radio frequency) noise. Impedance is the opposition to the flow of alternating current (AC). A ferrite choke is a passive electronic component that consists of a ferrite core (a ceramic material with magnetic properties) with a wire coiled around it, or sometimes simply a ferrite bead clamped around a wire. When RF noise travels along a cable, the ferrite choke's core absorbs the RF energy and increases the cable's impedance at high frequencies. This higher impedance makes it difficult for the RF noise to pass through the cable and forces it to flow through the ferrite core instead. The ferrite core then dissipates this RF noise as heat due to its lossy nature at high frequencies.

(24) (C) Gate.

In a Field Effect Transistor, the gate is the main control element. The voltage applied to the gate regulates the flow of current between the drain and source, making FETs voltage-controlled devices.

(25) (A) The resonant frequency of an LC oscillator is directly proportional to the square root of the product of inductance (L) and capacitance (C).

The resonant frequency of an LC oscillator is inversely proportional to the square root of the product of inductance (L) and capacitance (C). An LC oscillator is a circuit that generates an oscillating signal at a specific frequency, known as the resonant frequency. This frequency is determined by the inductance (L) and capacitance (C) in the circuit. The formula for calculating the resonant frequency (f) is $f = 1 \div [2\pi\sqrt{(LC)}]$. Increasing either L or C from the formula will decrease the resonant frequency and vice versa. This inverse relationship is why the statement is incorrect. Linear amplifiers are designed to amplify the input signal without introducing significant distortion, thus preserving the waveform shape. Class C amplifiers are highly efficient but introduce considerable distortion. This distortion is less critical for FM signals, as the information is encoded in the frequency variations rather than the amplitude. Amplitude Modulation (AM) relies on the faithful reproduction of the input signal's amplitude. The high distortion introduced by Class C amplifiers will severely degrade the quality of an AM signal and make them unsuitable for this modulation scheme.

(26) (C) The output signal power is less than half the input power at the cutoff frequency.

A low-pass filter allows frequencies below a specific cutoff frequency to pass through relatively unattenuated while attenuating frequencies above the cutoff. At the cutoff frequency, the output signal power is reduced to exactly half the input power. This corresponds to a -3 dB reduction in power or a 0.707 reduction in voltage amplitude. This fundamental characteristic of low-pass filters is often called the half-power point or the -3 dB point. Although low-pass filters significantly attenuate frequencies above the cutoff, they do not completely block them. There is always some residual attenuation in the stopband, which is the frequency range above the cutoff. Ideal low-pass filters do not amplify signals. They are commonly used to remove high-frequency noise from audio signals because most of the relevant audio information is in the lower frequencies.

(27) (A) A method of varying the frequency of a carrier to transmit digital data.

Frequency Shift Keying is a digital modulation technique where the frequency of the carrier signal is varied between discrete values to represent binary data. Each frequency shift corresponds to either a binary "0" or "1," allowing data to be transmitted digitally.

(28) (D) 28 kHz.

Carson's Rule provides a good estimate of the bandwidth of an FM signal: *Bandwidth (BW) = 2 × (Maximum Deviation + Highest Audio Modulation Frequency).*

In this case:

- Maximum Deviation = 10 kHz.
- Highest Audio Modulation Frequency = 4 kHz.

Substituting the values:

- BW = 2 × (10 kHz + 4 kHz) = 28 kHz

Therefore, the total bandwidth of the FM signal is 28 kHz.

(29) (C) The two distinct frequencies represent binary 1 and 0.

In Frequency Shift Keying (FSK), the terms *mark* and *space* represent the two distinct frequencies used to represent binary 1 and 0. FSK is a digital modulation scheme where the frequency of a carrier wave is shifted between two predetermined frequencies to convey information. The higher frequency is usually called mark, and the lower frequency is called space.

(30) (B) During decreases in signal frequency.

Coaxial cable attenuation is the loss of signal strength as it travels through the cable. This loss is influenced by several factors, with the signal frequency being a major one. Higher frequencies generally experience greater attenuation due to two primary factors: skin effect and dielectric loss. The skin effect causes the current to flow more toward the conductor's surface at higher frequencies, reduces the conductive area, and increases resistance. This increases power loss. The dielectric material within the coaxial cable also absorbs some energy from the signal, and this dielectric loss increases with frequency. Therefore, as the signal frequency decreases, the impact of both the skin effect and dielectric loss is reduced, which lowers overall attenuation in the coaxial cable.

(31) (B) 75 ohms.

A half-wave dipole antenna in free space has an impedance of approximately 75 ohms. This impedance is optimal for power transfer when matched correctly to transmission lines and equipment designed for similar impedance levels.

(32) (C) It consists of a shorted transmission line stub placed at the feed point.

A beta match is an impedance matching technique that connects a shorted transmission line stub in parallel with the driven element of the antenna at the feed point. This stub is usually less than a quarter-wavelength long and is shorted at the far end. The length of the stub determines the amount of inductance or capacitance introduced into the circuit. By carefully adjusting the length and position of the stub, the impedance presented at the feed point can be matched to the characteristic impedance of the coaxial cable (usually 50 ohms). A quarter-wavelength transformer is a different impedance matching technique, which involves a section of transmission that is line one-quarter wavelength long. Variable capacitors are sometimes used in conjunction with other matching techniques, but they are not the defining feature of a beta match. A beta match is an impedance-matching network, not a type of feedline. It can be used with both balanced and unbalanced feedlines.

(33) (D) Log-periodic antennas have narrow bandwidths.

Log-periodic antennas are known for their wide bandwidths, which is the range of frequencies over which the antenna can operate effectively. This characteristic makes them suitable for applications where a single antenna needs to cover multiple frequency bands. The logarithmic progression of element spacing in a log-periodic antenna ensures it maintains its performance characteristics across a wide range of frequencies. Harmonic frequencies are multiples of the fundamental operating frequency. Harmonic rejection is the ability of a transmitter to suppress these unwanted harmonics, which can cause interference to other radio users. Multiband antennas are designed to operate on multiple frequency bands. Due to the design compromises needed for multiband operation, they often have difficulty rejecting harmonics. In an inverted V dipole, a single mast or support structure holds up the apex of the V shape.

(34) (C) Increasing distance from the source.

Exposure to RF energy decreases significantly with increased distance from the source due to the inverse square law, which states that the intensity of RF energy diminishes as the distance from the source increases. This is an effective method to reduce potential exposure to RF energy.

(35) (C) To safeguard people and property from electrical hazards by providing wiring and installation standards.

The National Electrical Code is a set of standards for the safe installation of electrical wiring and equipment in the United States. It covers various topics, like wiring methods, materials, overcurrent protection, safety devices, and grounding. The primary goal of the NEC is to protect people and property from electrical hazards by establishing safe practices for electrical installations.

Download the Bonus Audiobook!

Scan the QR code below with your phone and you will be given a link to Google Drive where you can download the file.

If the QR code does not work for you, please contact us at info@newstonetestprep.com and mention the title of this book.

Made in United States
North Haven, CT
05 November 2026

21768647R00176